A HISTORY OF TECHNOLOGY AND ENVIRONMENT

This book provides an accessible overview of the ways that key areas of technology have impacted global ecosystems and natural communities. It offers a new way of thinking about the overall origins of environmental problems. Combining approaches drawn from environmental biology and the history of science and technology, it describes the motivations behind many technical advances and the settings in which they occurred, before tracing their ultimate environmental impacts. Four broad areas of human activity are described:

- over-harvesting of natural resources using the examples of hunting, fishing and freshwater use;
- farming, population, land use, and migration;
- discovery, synthesis and use of manufactured chemicals; and
- development of sources of artificial energy and the widespread pollution caused by power generation and energy use.

These innovations have been driven by various forces, but in most cases new technologies have emerged out of fascinating, psychologically rich, human experiences. This book provides an introduction to these complex developments and will be essential reading for students of science, technology and society, environmental history, and the history of science and technology.

Edward L. Golding is a Senior Lecturer at the University of Massachusetts at Amherst.

A HISTORY OF TECHNOLOGY AND ENVIRONMENT

From stone tools to ecological crisis

Edward L. Golding

Routledge
Taylor & Francis Group

LONDON AND NEW YORK

First published 2017
by Routledge
2 Park Square, Milton Park, Abingdon, Oxon OX14 4RN

and by Routledge
711 Third Avenue, New York, NY 10017

Routledge is an imprint of the Taylor & Francis Group, an informa business

© 2017 Edward L. Golding

British Library Cataloguing-in-Publication Data
A catalogue record for this book is available from the British Library

Library of Congress Cataloging-in-Publication Data
Names: Golding, Edward J., 1944- author.
Title: A history of technology and environment : from stone tools to ecological crisis / by Edward Golding.
Description: Milton Park, Abingdon, Oxon ; New York, NY : Routledge, 2017.
Identifiers: LCCN 2016028222 | ISBN 978-1-138-68585-7 (hbk) | ISBN 978-1-138-68586-4 (pbk) | ISBN 978-1-315-54295-9 (ebk)
Subjects: LCSH: Technology—Social aspects. | Technology—Environmental aspects. | Technology and civilization.
Classification: LCC T14.5 .G635 2017 | DDC 609—dc23LC record available at https://lccn.loc.gov/2016028222

ISBN: 978-1-138-68585-7 (hbk)
ISBN: 978-1-138-68586-4 (pbk)
ISBN: 978-1-315-54295-9 (ebk)

Typeset in Bembo
by FiSH Books Ltd, Enfield

To all the wild creatures—innocent victims of human desires

CONTENTS

PREFACE

The human story is one of the most remarkable chapters in the history of life. Beginning millions of years ago with tiny numbers and only the simplest of home-made tools, we have developed into a dominant population of billions possessing an incredible array of complex and powerful technologies.

The first tools used by human ancestors consisted of simple implements made of stone, bone or wood. By the time of the first farmers, 10,000 years ago, however, both the range of tools and materials in use and our overall know-how had expanded to include fire, pottery, bows and arrows, tool handles, sewing and weaving equipment, ropes and nets, baskets, musical instruments, boats, and on and on. Eventually, people had learned to do so many things with so many different kinds of tools and materials that a new concept was needed, one that combined the notions of the tools, materials and accessories used with that of the knowledge and skill of what to make and how to make it. Thus was born the concept of technology, a term whose Greek roots might reasonably be translated as the knowledge of skill. Some modern devices—think of an airplane or a computer—are easy to associate with the idea of technology. But people have invented so many different things, most of which scarcely resemble machines—think of a hand-held sewing needle, an ultra-microscopic synthetic chemical molecule, or a sewer pipe—that it is easy to forget how vast, pervasive and transformative the full range of human technologies has become.

Beginning in the seventeenth century, the rise of modern science created a basis for many new technologies which made it even easier to acquire natural resources and eventually to increase human numbers. By the nineteenth century, on the other hand, scientific progress also started to add a new dimension to our understanding of the earth and human origins. Geology and evolutionary biology began to outline not only earth's long, living history but also the concrete processes of natural selection that had shaped all creatures, including ourselves. Many people, to be sure, found these discoveries very frightening, and down to the present day these truths have frequently been attacked and rejected both by elements of organized religion and by many others who simply find them an affront to their sense of dignity or self-importance.

By the middle of the twentieth century, popular understanding of earth's history and the evolution of life began to gain wider acceptance. In the decades that followed, developing branches of science like ecology also managed to document both the actual functioning of natural communities and the ways in which human-induced changes were altering and disrupting them. Out of this emerged a new environmental world view, one centered on the idea that human civilization ought to be sustainable within the limits of what the earth could readily provide. Ever since then, this new outlook has vied with earlier mindsets which either reject scientific explanation entirely or insist that resource opportunities must be seized wherever they are available without regard to their impact on the environment. These struggles continue today and are likely to go on for a long time as advocates of these opposing perspectives seek to influence the options chosen by societies at all levels—personal, group, community, corporate, national and international.

Not all technologies turned out to cause damage to the rest of the living world, and even among those that did, there was a wide spectrum of effects ranging from the nearly benign to the extremely destructive. The focus of this book is on the development of several key areas of technology that have caused significant environmental damage: over-harvesting of natural resources; human land conversion, population growth, and migration; manufacture of large quantities of natural and synthetic chemicals; and creation and use of large amounts of energy and power. I hope that anyone who reads this book will find it easier to understand not only the background to our current environmental problems, but also some of the steps we can take to actually solve them.

ACKNOWLEDGEMENTS

I am grateful for the excellent education I received at Columbia and Yale Universities, which gave me a foundation that has allowed me to continue to study and learn right up to the present. I also learned a tremendous amount from my students at the University of Massachusetts Amherst. Many other individuals have also been important sources of information and of help, and three in particular stand out as providing critical assistance in my efforts to complete this project: Joseph B. Solodow of Yale University, Karen E. Stothert of The University of Texas San Antonio, and my wife, Linda Root Golding. All three offered me wonderful encouragement and invaluable advice during the long process of turning this manuscript into a book. In addition, I would like to thank Guy Lanza of the State University of New York at Syracuse for his helpful review of several chapters. Besides these individuals, I would in addition like to recognize the very helpful institutional support I received from the University Library and the Office of Information Technology at the University of Massachusetts Amherst, and also from the Goodwin Memorial Library in Hadley, Massachusetts.

1

OVER-HARVESTING OF NATURAL RESOURCES

Hunting and Fishing

Introduction

Among the important human activities which have placed us on a collision course with the rest of the living world, the over-harvesting of natural resources is one of the most obvious. For hundreds of thousands of years, our ancestors' tiny numbers and simple technologies posed little threat to the supply of most of the useful things that nature and the earth could provide. By around 50,000 years ago, however, following the development of language, advanced tool-making, and complex cooperative behavior, things seem to have changed. Starting with the over-hunting of large game animals, people began to gradually reduce or even completely exhaust many different natural resources: big game animals, high grade deposits of valuable minerals, quality timbers, different kinds of animals and plants, fresh water, and even clean air.[1] The greatest drivers of these changes were our increasing numbers, our geographical expansion out of the Old World, and the on-going march of technology. Population growth created demand for ever larger amounts of more and more varied goods; technological innovation led to ingenious new methods of capturing resources as well as to seemingly endless ways of incorporating them into economically valuable products. Throughout history, numerous natural resources were over-exploited in this way, and many scholars have documented the reduction or exhaustion of one valuable material or another.[2] In our account, we will limit ourselves to just two examples, the excessive harvesting of wild animals, and the over-use of fresh water.

Tools and Hunting

Many anthropologists believe that even before intentional stone tool-making began, our pre-human ancestors used stones as pounders, choppers and scrapers, along with sharpened sticks for digging and foraging. There may also have been a slow progression from using natural stones as hammers—for driving a digging stick into the ground or for attacking prey or enemies—to hammering one stone against another to make a sharp edge. There is evidence that stones with edges were being created by around 3.4 million years ago, in what is today Ethiopia, in order to scrape the meat off animal bones.[3] Stone tool-making marked an important turning point in human evolution, one with profound implications for the rest of the living world. Once it began, natural selection led to changes in our ancestors' bodies which allowed them to more easily hold sticks and stones, and to manipulate them in ever more complex ways. Changes occurred in the arms, hands and fingers, along with less obvious modifications in the brain and nervous system with the result that our ancestors developed remarkable manual dexterity and hand–eye coordination.[4]

The lack of abundant fossil material makes it difficult to tell how widespread tool use by human ancestors actually was after its first occurrence in Ethiopia— early humans were not very numerous and tools made of wood or bone tended to decompose over time. Stone, however, is one of earth's most durable materials, and we have a better record of early humans' stone artifacts than we do even of their bones. (In contrast, the archaeological record of our ancestors' early use of fire is much sparser and harder to interpret, leading to considerable controversy about just how early this other key *tool* of early humans came into use. Some researchers see evidence of it dating back more than a million years, while others believe that the use of controlled fire began only a few hundred thousand years ago.)

By about 2.6 million years ago, stone tools seem to have become more common, with the oldest ones from that time being recovered from Ethiopia. Many, however, have also been found in or near Olduvai Gorge in Kenya, a location where many key human fossils have also been unearthed. This region also provided a name for this early style of stone tool—*Olduwan*—of Olduvai. For roughly the following million years, until about 1.65 million years ago, Olduwan-style choppers, cutters and scrapers seem to have been the only stone tools our ancestors made and used. Generally, these implements are of two types: larger stones that have been chipped to make one or more sharp edges, and which were probably used as axes and choppers; and smaller, sharp flakes that had been knocked off the larger stones and were probably used as knives and scrapers. So far there is no direct evidence that these very early stone tools were used as hunting weapons, although they may well have been used to sharpen wooden spears or to cut up animal carcasses.

Most people today have little or no experience of stone tools beyond the occasional use of a natural rock as a primitive hammer. But in the past this was not the case. Human evolution over several million years was closely tied to the creation and use of carefully crafted stone implements. To shape cutters—partic- ularly the more complex blades that started to be made after 1.65 million years ago—our ancestors had to evolve not only dexterity and fine motor control in their hands and arms, but also the mental capacity to imagine both the variety of stone blades that could be made, and the complex steps required to make them. For this reason, stone tool-making not only marked the beginnings of human technology, but also likely served as a powerful spur to the evolution of human intelligence and creativity. Our own species, *Homo sapiens*, for example, did not appear in Africa until roughly 200,000 years ago. But even when it first arose, it was heir to an already ancient tradition of stone tool manufacture, a set of skills that had allowed earlier humans without large claws or teeth to efficiently hunt, gather and process food in the face of competition from more heavily toothed and clawed carnivores and scavengers.[5]

Numerous anthropologists have spent their working lives studying prehistoric stone tools and the people who made them. Unearthed on every inhabited continent, many tens of thousands of such objects have made their way into museums and other collections, resulting in numerous detailed histories of various stone tool *industries* from different parts of the world. Around 1.65 million years ago, for example, there started to be widespread evidence of an *Acheulean* stone industry (named for Saint-Acheul in northern France) in which early humans, first in Africa, and then in Asia and Europe, made sophisticated pear and oval-shaped hand axes with stone blades symmetrically shaped on both sides. Acheulian stone tool-making was in fact the longest lasting of all human technologies, and it continued for more than one million years until roughly 250,000 years ago.[6]

In the earliest days of tool-making our ancestors seem to have lived by a combination of gathering plant food, scavenging dead animals—perhaps by driving other predators away from carcasses after they had made a kill—and by hunting small, easy-to-kill prey, probably with wooden clubs. It seems likely that the majority of early stone tools were used to cut up animal and plant food. As time went on however—and the archaeological record here is poor—stone blades also came to be combined with wooden handles to make hatchets, axes and spears, the last of which marked a particularly important advance in human weaponry. Attached to a long wooden shaft, a sharp stone blade could now be used to attack prey animals or to defend against predators while keeping the hunter some distance away from the animal's claws, hooves or teeth. This made big game hunting a feasible proposition, if still a very dangerous one. We do not know exactly when this began, but by 400,000 years ago in Germany there are important archaeo-logical remains of a site occupied by humans, likely *Homo heidelbergensis*, that contain large, well-crafted wooden hunting spears, roughly seven feet long, which seem to have been designed to throw at prey animals. They were discovered near the town of Schoeningen, adjacent to a coal mine, in what were originally shallow lake deposits. Rapid burial there in sediments lacking oxygen seems to have allowed the wood to be preserved. The site also included stone blades made of flint, and a wooden handle that was notched to hold those blades; with them were the remains of butchered horses. Additional evidence of the hunting success of the spear-makers was found nearby in the form of numerous skeletal components of red deer, bison and a variety of smaller animals.[7]

In Europe by 250,000 years ago, Neanderthals (*Homo neanderthalensis*) had begun to make even more sophisticated edge tools using a complex procedure known as the *Levallois technique*. This method involved carefully preparing a large, rough stone by rounds of chipping, followed by a single, precisely placed blow which then produced an almost completely formed point. It was a tricky and intricate process requiring careful planning and good design sense in addition to considerable physical strength and hand–eye coordination.[8]

The technological developments we have been describing were occurring in the Old World during the Pleistocene epoch, a time characterized by repeated episodes of widespread glaciation in Eurasia and North America as well as in high mountains all over the world. Such large climatic changes were themselves stressful for animal and plant populations, and it is against this background that the impacts of changing human hunting technologies should be viewed. It is also important to remember that the two and a half million years of the Pleistocene were not uniformly cold and ice-covered—there were a number of warmer periods when the ice retreated, alternating with a number of major and minor glacial advances. Overall though, it was still a time of great climatic stress in most of the cooler and higher altitude parts of the world, with large glaciated areas being largely denuded of their vegetation, animals and people.[9]

Advance and retreat of the ice sheets did not happen all at once; in the Midwestern USA, for example, the last Pleistocene ice sheet is estimated to have advanced roughly 200 feet per year, or about one mile every twenty-five years.[10] In response, many kinds of plant communities, and most animals, were able to move south as the glaciers expanded, and then north again after they receded. Exceptions occurred in places where barriers like high mountains lay to the south, as the Alps did in central Europe. In such cases, lacking suitable habitat to move south towards, some species became extinct. During this time, humans did best in warmer, ice-free lowlands in the Old World, where they continued to invent new kinds of weapons and tools as they increasingly became important predators of large animals.

By 200,000 years ago in Africa the first *anatomically modern* humans, members of our own species, had appeared; it is possible that they already possessed some aspects of culture and language including the earliest art, music and religion. By 100,000 years ago, archaeological remains show continuing cultural advances including intentional burial of the dead and more uniform, precisely fashioned stone tools. By 60,000 years ago, our immediate ancestors had also begun to make small, sharp stone points that look like arrowheads, suggesting that their hunting equipment had further diversified to include bows and arrows. Also within the last 50,000 years, prehistoric peoples invented a new way of making sharp stone tools by grinding rather than chipping, an approach that allowed them to use additional kinds of stones as starting materials; the first evidence of this comes from northern Australia about 35,000 years ago.[11] Axe heads prepared by grinding were smoother than chipped axes and hence better at heavy wood cutting. For the first time, this made it relatively straightforward for people to intentionally chop down medium and large trees. This increased their supplies of wood for fuel, building and tool-making while also raising the level of potential damage that people could do to forested ecosystems. Fire had probably already been used to alter forests for tens or even hundreds of thousands of years. Now, however, a combination of chopping, fire, and later on the grazing of domestic animals meant that it became easier and easier for early people to alter or even eliminate forests, reducing their populations of animals and plants or even wiping out some species entirely.[12]

The Spread of Modern Humans and the Impact on Natural Communities

Between 60,000 and 120,000 years ago, our own species, *Homo sapiens*, began to expand its range outwards from its original homeland in Africa. By that time, Africa had been inhabited by various human species for more than two million years, and Europe and Asia for more than a million and a half years. On those continents some large animals seem to have been driven extinct by human activities. In Africa, these included several species of elephants, a giant hippopotamus, various warthogs, giant buffalo and antelope, a wolf and several saber-toothed cats. To some extent, however, Old World faunas' long existence side by side with slowly advancing human hunting techniques and weapons may have allowed them to adapt in ways that reduced species' extinctions.[13] There is also some evidence that in the Old World, modern *Homo sapiens* was largely confined to temperate areas, while the largest mammals like mammoths, that survived until ten to twenty thousand years, were mostly inhabitants of frigid Arctic zones.[14]

Once they left Africa, modern *Homo sapiens* are believed to have moved initially north and east across the Red Sea, around Arabia and over towards India; later on, they also moved north and west into Europe and central Asia where Neanderthals had long been present. We do not know what pushed them to migrate—reduction of game populations in their old lands, escape from predators and parasites, local overpopulation, or restlessness and curiosity. By 70,000 years ago, they had become highly skilled makers of stone tipped weapons and tools and also had full control of fire, which they knew how to use for cooking, warmth, defense, hunting, processing of materials *and* habitat modification. In addition, they possessed language and advanced forms of cooperation and team-work. Together, this package of culture and technology made them formidable hunters, able to attack even very large game with their stone and bone-tipped weapons. Around the same time, they also developed some kind of watercraft—most likely rafts—which allowed them to cross bodies of water too wide to wade or swim, and thus to enter new lands that earlier species of human had never reached. Equipped with these, they made their way down from India across the numerous small and large islands of the East Indies, until finally, by about 50,000 years ago, they reached Australia, an entire continent that had never before experienced human presence.[15]

When humans arrived, Australia had been isolated from Africa and Eurasia by water barriers for tens of millions of years, and the first settlers encountered a wide array of animals which had never before coexisted with people or even with other primates. The fauna of pre-contact Australia included numerous remarkable creatures, particularly marsupial mammals, some of which weighed as much as two tons. Surviving today only as skeletal remains, they are known as the extinct *Australian megafauna*. Among them were several very large relatives of the small surviving Australian wombat. One was the *Diprotodon*, a creature the size of a rhinoceros, while another wombat relative superficially resembled a modern tapir in size and shape. A third extinct species was a fierce-looking predator, *Thylacoleo carnifex*, a *marsupial lion* which grew to the size of a leopard and weighed over 300 pounds (136 kg). Many kinds of giant kangaroos and wallabies were also present, some standing 8–10 feet tall and weighing 500 pounds (226 kg). There was also an *echidna* or spiny anteater the size of a sheep—the surviving echidna is a rabbit sized, egg-laying mammal related to the platypus. In addition, there were some enormous reptiles including a giant monitor lizard which grew to more than 20 feet in length and may have weighed over 4,000 pounds (1814 kg). All in all, roughly forty species of giant land vertebrates were present in Australia when people first arrived. But by around 47,000 years ago, after only a few thousand years of human settlement, they were nearly all gone. Their loss resulted from a combination of impacts, particularly direct hunting coupled with intentional burning of the environment in order to encourage the growth of grass on which smaller (and more rapidly reproducing) wallabies and kangaroos could feed. As a result, Australia became the first large area of the earth to experience catastrophic ecological disruption in the wake of initial human settlement.[16]

Killing large animals was very appealing to hunters because each individual taken provided a great deal of food, and there may well have also been status competition among men trying to kill the largest and most impressive prey animals, just as there is today among *trophy* hunters. Prior to human hunting, however, many unusually large animals had been relatively free of natural enemies, and evolution had often responded to this by adjusting their reproductive rate downwards so that they would not waste resources by producing extra young that the environment could not support. (As a generalization, larger mammals live longer than smaller ones, but also reproduce more slowly. African elephants, for example, do not breed until around age thirteen, are pregnant for almost two years, and generally have just one baby every four to five years. In contrast, a small mammal like the European rabbit can breed by the time it is four months old, have up to seven litters per year, and produce forty offspring annually.) Low birth rates likely placed megafauna species in particular danger when suddenly confronted with clever humans armed with stone-tipped weapons and employing sophisticated, cooperative hunting strategies. The result in many cases was over-hunting, the decline of prey species populations, and eventual total extinction.[17]

The demise of the individual species involved was a great loss to the living richness of Australia, but the environmental damage was far more extensive even than what might be expected from the number of species lost. The majority of megafauna species were plant eaters whose feeding was a central part of the basic functioning of the ecosystems in which they lived. Their activities helped establish the living balance among plant species and the cycling of energy and nutrients, and thus helped to create characteristic ecosystem patterns which had developed slowly over many thousands of years. They ate so much, and cropped vegetation so heavily that their extinction shifted growth and competition among plants and the other creatures that fed on them, and this, in turn, probably caused other species to go extinct.[18] Even today, for example, African elephants consume more than 300 pounds (136 kg) of plant food daily, and in this way play a key role in maintaining the structure of their natural communities.[19]

In addition, once human hunting and burning had wiped out most or all of the larger Australian herbivores, the few large Australian carnivores that had been able to prey on the mega-herbivores (or at least, on their young) also suffered declines, eventually to the point of extinction. This was the pattern in Australia between forty-five and fifty thousand years ago, and it was to be repeated again and again wherever people entered new lands containing large mammals, birds and reptiles. Even in Europe and Asia, where *Homo erectus* and *Homo neanderthalensis* had hunted for hundreds of thousands of years before *Homo sapiens* eventually took over, by 12,000 years ago improved weapons and hunting techniques, combined with human population growth and possibly glacial climate change, had contributed to the extinction of much of the local surviving *megafauna*, including mammoths, Irish elk, woolly rhinoceros, steppe bison, wild aurochs (the ancestor of domestic cattle), cave bear, and others.[20]

After Australia, the Americas were the next large land masses reached by humans as our range expanded. The exact date is uncertain, but people had definitely arrived in North (and probably also South) America by 15,000 years ago. Before that time, those two New World continents had some of the richest communities of large mammals on earth. Most of these had evolved in the Americas, and to these had been added several large species which had entered North America from Asia across a Bering land bridge roughly 40,000 years ago, most notably mammoths, mastodons and the American bison. With the exception of these new arrivals, none of the New World species had ever before encountered human hunters, and as in Australia, our meat-hunting ancestors, armed with stone pointed weapons and sophisticated, cooperative hunting strategies, were able to have a significant impact on North and South America's larger species. By 10,000 years ago, North America was almost completely emptied of megafauna, having lost more than *three dozen* species of large mammal, including horses, camels, llamas, deer, pronghorns, oxen, moose, bison, giant beaver, several ground sloths, a bear, and three kinds of elephants. Also gone were several large predators including a saber tooth cat, two lions, a cheetah and a wolf. And South America fared little better, losing at least *two dozen* large mammals including elephants, camels, sloths, saber tooth cats, a bear, numerous large ground sloths, horses, giant armadillo-like animals, and *Toxodon*, a large rhinoceros-like plant-eater with a head resembling a hippopotamus. (Many scholars support this account of the human role in the extinction of many of earth's largest mammals between ten and fifty thousand years ago. Some, however, argue that climatic change, which also occurred in the same period at least in Eurasia and the Americas, was the main cause of megafauna extinction; others suggest that the arrival of humans, dogs and other domestic animals may have introduced extremely virulent diseases which were responsible for the mass die-offs.[21])

Following the extinction of so many large animals, first in Australia and then in North and South America, surviving large species continued to disappear following human arrival in new lands right up to recent times. Many larger Mediterranean islands, for example, had evolved unique species of mammals such as dwarf hippos and elephants. But by 9,000 years ago, places like Sicily, Crete, Majorca, Minorca and Malta had lost all of their larger animals in the wake of human settlement, although it is not yet clear whether direct hunting, habitat modification or introduction of exotic species was the most important causes of these disappearances. Between five and ten thousand years ago, the same thing happened on islands in the Caribbean, where sloths and other large creatures seem to have been hunted to extinction. Madagascar, a large, tropical island in the Indian Ocean east of Africa was first settled by people roughly 2,000 years ago, and since then it has lost many unique large animals including a giant tortoise, three native species of hippos, seventeen of its fifty species of lemurs (primates like us), some as large as a gorilla, and several species of giant, flightless elephant birds.[22]

Recent Extinctions

The last significant land areas reached by people other than frozen Antarctica were the two large islands of New Zealand, first reached by Polynesians roughly one 1,000 years ago. Geographically isolated, New Zealand originally had no native mammals at all except for a few small species of bats whose ancestors had flown in. Instead, all of the large browsing and grazing animals were flightless birds known as moas. There were nine species, with the largest standing more than ten feet tall and weighing over 500 pounds (226 kg). Not long after Polynesians arrived, as settlers increased in numbers and gradually moved south, one population of moas after another was intensively hunted to the point of complete extinction. In just a few hundred years—by around 1400—the last of them was gone, followed by Haast's eagle, the largest eagle ever known, and the only native New Zealand predator large enough to have preyed on the larger moas.[23]

Closer to the present, the pattern was similar. The Steller's sea cow, for example, was a large marine mammal related to tropical manatees and dugongs. (The latter still survive, but are declining and endangered due to human activities.) It inhabited cold waters of the Northern Pacific, and its ancestors had lived successfully in the oceans for millions of years. Fossil evidence shows that it once occupied a wide range of coastal areas right around the Pacific from Japan to California, and that it had been eliminated by hunting from nearly all that range even before written records were kept. Its first scientific description was made in 1741 by Georg Steller, the naturalist on Vitus Bering's North Pacific expedition for czarist Russia. Bering's party traveled from Kamchatka in Siberia eastward along the Aleutian Island chain, eventually setting up small settlements in what is today Alaska. Steller encountered the last sea cows in the uninhabited Commander Islands, east of Kamchatka, where a population of a thousand or more had managed to survive. They were large, peaceful, plant-eaters, growing 25 to 30 feet long, weighing several tons and subsisting mostly on kelp. Lacking natural enemies other than humans, they were adapted to slow reproduction, and they were easy for sailors to capture for meat, oil (rendered from their fat) and skins. By 1741—only twenty-seven years after they were first recorded—the last individuals on earth had been killed by hunters.[24]

Such losses continued and even gathered pace in the 1800s as human populations grew larger, boats and land transportation improved, and the introduction of modern animal traps and firearms made hunting and collecting ever more efficient and deadly. It was also an era when regional and global trade by ship expanded significantly (see Chapter 3), with the result that *market hunting* became more common, dramatically increasing the potential demand for harvested animals whose products could be transported to ever larger numbers of customers.

The great auk, for example—the "penguin of the north"—was the largest member of the auk family, a group which also includes birds like puffins and guillemots. It was a large, flightless seabird, weighing up to 11 pounds (5 kg) that was native to the North Atlantic and bred on off-shore islands ranging from the Canadian Maritimes to Greenland, Iceland and the northwestern British Isles. In the water, it was highly skilled at catching fish; on land, where it usually went only to breed, it was relatively awkward and easy to prey upon. Like many larger birds and mammals, it also had a relatively low reproductive rate, laying just a single egg each year. Its natural enemies were uncommon—killer whales, sea eagles and polar bears took a few. Not being heavily preyed upon, it tended to be quite trusting when it encountered human hunters. Its original geographical distribution all around the North Atlantic was reminiscent of that of Steller's sea cow in the North Pacific; remains from archaeological sites show that even 100,000 years ago the great auk was being eaten by Neanderthal humans, and the species was also depicted in cave art produced by *Homo sapiens* as long as 35,000 years ago.[25]

Before human hunting and egg collecting, there are thought to have been millions of great auks around the North Atlantic. In historic times, remaining populations were heavily preyed upon by people in the outer British Isles, in Iceland, and on islands in the northwest Atlantic, where eggs, feathers and meat were collected. Sometimes the meat was used for fishing bait. By the 1500s, populations in Europe had mostly been wiped out due to the demand for feather down, while scattered colonies were still found in the northern and northwestern Atlantic. Around that time, the vast expansion of fishing off maritime Canada and New England started to reduce New World populations, as sailors intensively collected both birds and eggs. By the late 1700s, the decline of the species had been noticed, and museums and private collectors seeking eggs for their egg collections—a popular upper class hobby at the time—further reduced the auk's fast dwindling numbers. By the late 1830s, perhaps only one hundred auks were left, mostly near Iceland. Despite this, and even in the face of protective legislation that had been passed by that time, egg collecting and hunting continued. The result populations were decimated, mostly for meat, but to some extent also for decorative feathers for the millinery trade.[29]

In 1825, the world's first public railroad was inaugurated in Britain, and over the next two decades, early rail lines also began to be laid in the United States. By the 1850s, many places in the eastern United States were connected by railways, and by the early 1860s the American Civil War had given added impetus to railroad construction. Once the war was over in 1865, the building of a transcontinental railroad connecting the east and west coasts became an important national priority. By 1869 a line stretching east from California connected with a line coming west through the Rocky Mountains to create a continent-spanning route.[30] Railroads had a profound impact on society; they transformed the United States by making it possible to rapidly move people, goods and information hundreds or even thousands of miles at relatively low cost, and this, in turn, spurred both economic growth and social development. This progress, however, did not come without a substantial environmental cost. Railroads caused significant pollution and required the mining of coal or the clear-cutting of forests to provide their fuel. And by helping to create regional and national markets for products, they also created conditions which made the mass slaughter of wildlife economically profitable by allowing enormous quantities of meat and animal products to be rapidly and inexpensively shipped over long distances. In response, a national *market hunting* industry grew up dedicated to killing wildlife in large quantities for sale in distant markets. And unlike traditional local and small-scale regional hunting, in which residents were often committed to maintaining the breeding stocks of wildlife to ensure future availability, market hunters viewed wildlife purely as a commodity, as something whose sole value lay in the immediate cash price it could bring in the market place.

In addition to a revolution in transportation created by the development of railways and steamships in the first half of the nineteenth century, there was also a dramatic change in rapid, long-distance communications with the invention of the electric telegraph. As we will see in Chapter 6, each decade of the early nineteenth century brought important advances in electrical technology, and by the 1830s several experimenters were at work on a variety of systems using copper wires to rapidly send messages over long distances. By the mid-1840s, workable telegraphs were being produced, and the first commercial line in North American was put into service along a railroad right of way in Pennsylvania. The nearly instantaneous transmission of information provided by the telegraph might seem like an unlikely factor in the growth of market hunting, but by making it possible for commercial wildlife suppliers to learn the quantities of wildlife that distant markets were seeking, and the prices they were willing to pay, the telegraph too played a key role in the development of nineteenth-century market hunting.[31]

In addition to railroads and telegraphs, the 1800s also saw a large number of improvements to rifles and shotguns. Most early firearms had to be slowly and painstakingly loaded by ramming powder, bullet and packing down the muzzle, and then fired by having a spark or flame ignite the gunpowder. Such a system understandably worked poorly, if at all, in wet weather. Early in the nineteenth century, however, serviceable paper cartridges were developed containing the bullet, the gunpowder and a chemical that would ignite when struck with a needle. This eliminated the need for an external flame or spark, and permitted firing under damper conditions. Such cartridges were then combined with breech-loading mechanisms to produce more reliable single-shot firearms which could quickly be reloaded. By the 1850s, practical, lever-action, repeating rifles were also invented that fired bullets loaded into metal cartridges, a form of ammunition that could reliably be packed into a magazine. Such weapons made it possible to fire up to sixteen shots before reloading. By the mid-1880s, repeating shotguns were also developed, capable of firing five shells in a row. Throughout the nineteenth century, advances in precision manufacturing made it possible to produce both firearms and ammunition that allowed increasing accuracy. Taken together, all of these innovations provided hunters higher rates of fire in all kinds of weather and allowed them to kill more wildlife.[32]

The Passenger Pigeon

As wild ducks and geese came to be heavily hunted and sold commercially in the early nineteenth century, abundant land birds were also targeted. The most notable of these was a North American woodland bird, *Ectopistes migratorius*, commonly known as the passenger pigeon. It was a fairly large, multi-colored dove which found the unbroken deciduous woodlands of the eastern United States and Canada to be a very productive habitat. During their migrations, the birds would pass overhead in vast flocks for hours and even days at a time, and individual nesting colonies were often 40 miles long and several miles wide, containing hundreds of millions of birds. Throughout the period from the 1600s to the early 1800s, the passenger pigeon's total population is estimated to have been between two and five *billion* birds, making them the commonest species in North America and perhaps the most abundant in the world.[33]

Adult passenger pigeons were edible, and their large nestlings, or *squabs*, were considered a delicacy. Their abundance had made them a popular local source of game even in the 1700s, and by the 1800s, local commercial hunting made them the commonest meat provided to slaves, servants and the poor. Existing in such overwhelming masses, they were relatively unwary, as the enormity of their flocks meant that natural predators were never numerous enough to seriously affect their populations. Human hunters, however, were another story. Even before improved firearms were developed during the nineteenth century, pigeons were taken by the millions. Some were shot, some were baited with alcohol-soaked grain and then netted; many squabs were knocked from trees with long poles, and sometimes fire was used to scare off the adults and cause the young to fall out of the nest. Huge numbers—250,000 per year in the 1870s, when pigeon populations were already in sharp decline—were also captured solely to use as live targets in shooting galleries; they seemed to exist in inexhaustible numbers and were exploited without the least concern for their long-term survival. By the mid-nineteenth century, some observers noted that their populations were declining, and the relentless clearing of woodland for farms and pastures also reduced the pigeons' suitable habitat, although overall pigeon numbers were probably still several billion, mostly concentrated in upper Midwestern states like Michigan. In the 1850s, however, railroads were built into that part of the country, and large, commercial market hunting operations soon began rail shipments of pigeons to densely settled areas along the east coast, where the birds generally sold for a few cents each. In 1869, for example, more than seven million birds were captured and shipped east just from the single Michigan county of Van Buren. By 1890, this unsustainable onslaught was rapidly driving the species to extinction, at least in part because their evolved behavior required the stimulation of enormous flocks for successful mating and nesting. Small numbers of wild passenger pigeons continued to be observed during the 1890s, but no successful nesting flocks were left, and the last wild birds ever seen vanished in Ohio by 1900. Only a few captive birds remained, and in 1914 the very last passenger pigeon alive on planet earth died in a zoo in Cincinnati. After millions of years of successful survival on earth, through ice ages and other challenges, human over-harvesting drove the passenger pigeon from almost unbelievable abundance to complete extinction in just about one hundred years.[34]

The American Bison

Passenger pigeons were key components of the ecology of eastern North America, and natural systems will take a very long time to recover from the pigeons' demise. Even if they had been the only creatures to be driven to extinction, it would have represented a gigantic loss of natural richness that at a minimum had taken many tens of thousands of years to develop. Sadly, however, pigeons and water birds were not the only creatures blindly over-hunted in the nineteenth century. In North America another species on the endangered list was the continent's largest surviving land animal: the American bison or buffalo. This creature had survived thousands of years of hunting by native peoples, successfully making it through the holocaust of the megafauna, perhaps because it had a higher birth rate than other large, plant-eating mammals. Its vulnerability to native hunters did increase after Europeans introduced riding horses and firearms into North America, but tribal peoples generally put limits on their kills in order to sustain the buffalo populations. Bison had originally ranged over much of what today is the United States, central Canada and northern Mexico. In their heyday, they were the most common large mammal on earth and their North American populations may have amounted to as many as one hundred million. As herbivores they played a central role in the food webs, community life histories and nutrient flows of the ecosystems they dominated. They also formed the survival base of the Plains Indians who had learned to use nearly every part of the bison for one purpose or another.[35]

With such enormous numbers and large body size, the bison played a dominant ecological role over a vast natural area in interior North America. Like many other large land masses, North America shows a geographical pattern of having more rainfall near the coasts with generally drier conditions further inland. Seashore locations like New York City may receive as much as 50 inches of annual precipitation, an amount more than adequate to support the growth of forests. Relatively high levels of precipitation extend westward a considerable distance until their decline to below about 40 inches begins to create a mixture of woodland and grassland. Moisture, of course, is not the only determinant of the dominant vegetation; soil drainage is also important, as is the frequency of lightning-set summer wildfires and even the intensity of grazing. From the Mississippi River west all the way to the Rocky Mountains (and even in a few areas east of the river), much of the vegetation present before European settlement was grassland, the Great Plains biome where the bison was ecologically dominant.

As ecological understanding expanded during the twentieth century, particularly in the decades following the 1960s, the remarkable nature of grassland vegetation began to be understood, even after more than 95 percent of its former extent had already been converted to agriculture. Where preserved patches were studied, remarkable richness and ecological dynamics were found. On just the 11,000 acres (4451 hectares) of the Tall Grass Prairie National Preserve in the Flint Hills of Kansas, for example, fully 500 different kinds of plants have been identified, most of them broad-leaved herbs which grow along with grasses; high animal diversity is also present, and these figures offer a taste of what the original living richness of this region must have been like just a few centuries ago.[36]

Until the mid-nineteenth century, a very large portion of all the food and energy present on the Great Plains flowed through bison. The giant herds ate almost inconceivably large amounts of grass and broad-leaved plants, drank significant amounts of water, changed the chemistry and biology of the grasslands with their heavy manure and physically altered the ground with their hooves. Their bodies and their droppings provided habitats for numerous small arthropods and other invertebrates which, in turn, provided food for birds and some mammals. In addition, their constant cropping actually tended to *increase* the plant species diversity of the areas they frequented. They did this because in the absence of large-scale grazing by bison, a small percentage of the native grasses and other plants grew strongly enough to out-compete many neighboring plant species. With low to moderate levels of repeated grazing, however, overall species richness tended to be maintained.[37] Between the late nineteenth and the middle of the twentieth century, the slaughter and near extinction of the American bison was widely viewed as an example of the over-exploitation of natural resources and the endangerment of an iconic animal. It was only in the second half of the twentieth century, however, as a fuller understanding of the ecology of the American prairie began to emerge, that the true impact of the loss of the wild buffalo herds began to be grasped.

The bison's ultimate fate was closely intertwined with the history of European settlement of North America. After initial focus on locations along the ocean coasts and those of the Great Lakes, by the 1700s the growing settler populations had begun to push inland, establishing villages and towns west of the Appalachian

Mountains by 1800. As new settlements sprung up, Native Americans were commonly displaced, while areas that had been forest or prairie were increasingly converted to cultivated fields, pastures and grazing land (see Chapter 3). Most of the early expansion east of the Mississippi occurred in forests and woodlands that were not prime bison habitat. But by the 1820s and 1830s, European settlers were arriving in greater numbers around the Mississippi River, and before long they slowly began to extend their farms and ranches into the heartland of the Great Plains onto land that had been prime buffalo habitat. By this time, most Native Americans east of the Mississippi had either succumbed to introduced diseases or been forcibly relocated, but west of the river Indian communities were still reasonably intact and were dominated by nomadic and semi-nomadic tribes whose way of life revolved around the buffalo. Contact with Europeans had brought them firearms and horses, but even though they continually hunted bison, they seem to have done so in a sustainable way.

From 1830 until around 1860, encroachment on the territories of the Plains Indians and hence on the natural habitat of the American bison was gradual, and in 1860, the outbreak of America's Civil War slowed the pace of settlement for a few years. In 1862, however, while that war was still raging, the US Congress passed two pieces of legislation that drastically affected the Great Plains and the future of both its human and wild inhabitants. In May of that year came the Homestead Act which offered 160 acres of land on the western frontier at no cost to anyone who would settle there and farm the land for five years.[38] Then, in July, came the Pacific Railroad Act of 1862—"An Act to aid in the construction of a railroad and telegraph line from the Missouri river to the Pacific ocean, and to secure to the government the use of the same for postal, military, and other purposes."[39] This law, followed by additional legislation in 1863 specified uniform rail gauges for all lines and offered generous grants of land and low-cost loans funded by government bonds to builders of transcontinental rail lines. The lands involved, of course, were already occupied by many thousands of Native Americans as well as by tens of millions of buffalo, but the act was passed without consultation with the affected tribes and with no concern for protecting the natural communities.

After the end of the Civil War in 1865, larger number of settlers began to push west to take advantage of the lands offered by the Homestead Act, and this brought them into immediate conflict with both Indians and the free-ranging herds of buffalo. The US Government's response to these clashes was to deploy the US army—a force that had grown battle-hardened and more sophisticated during the Civil War—in order to *pacify* the frontier. Additional forts were soon built and more troops sent west in order to safeguard the European population. In addition, the conclusion of the Civil War led citizens and politicians in the victorious north to once again focus on business development and technological advancement, and soon the expansion of railroads across the Great Plains became a local and national priority.[40]

By the 1870s, many forces had come together to threaten the future of the bison, the most abundant large mammal to survive into modern times. And except for the Plains Indians, whose way of life was being destroyed, no one seems to have expressed much concern about the bison's future until it was just about annihilated. The Federal government and its military representatives were determined to destroy Native American resistance to the settlement of their lands, and removing the buffalo took away the Indians' most important source of food and raw materials. Farmers and ranchers wanted land free of wild grazing animals, and they also supported forcing the Indians off the land and on to reservations. For the new railroad lines, the wild grazing animals were both a nuisance and a hazard, as herds wandered freely over their tracks, sometimes blocking trains or even colliding with them. Business people wanted to make money by selling millions of buffalo hides and robes for which a strong market existed both in North America and in Europe. Factories in the eastern states and abroad sought vast quantities of extra-strong leather for use as machinery belting in their mills, a purpose for which bison skins were considered ideal.

For the bison, it was a perfect storm, and they died in their millions, mostly slaughtered for their skins, with their bodies left to rot on the grasslands, although in some cases meat was also harvested, transported and sold. Sometimes, small businessmen also went around later and collected their bones to use as fertilizer. In the 1870s, hundreds of buffalo hunters were continuously active, pushing farther and farther west. At first they followed the new railroad lines, but soon also branched out in between them. The best estimates are that thousands of buffalo were killed each and every day throughout the 1870s, with at least one hunter claiming to have personally slaughtered more than 20,000. No population of large animals, not even one originally numbering in the tens of millions, could long withstand this kind of pressure, and by the early 1890s the heretofore uncountable herds were reduced to the brink of extinction with only about 300 individuals remaining. At that point, when scarcely any were still present to protect, the bison received legal protection and several tiny wild herds were re-established, with a small, free-ranging group located in Yellowstone National Park in Montana. Today, the bison is no longer threatened with out-and-out extinction, although genetically speaking, its status is still very problematic. The vast majority of the hundreds of thousands that survive on ranches today no longer carry the full genome typical of their species, but instead have been hybridized to a greater or lesser extent through interbreeding with domestic cattle.[41]

The Eskimo Curlew

As we have seen, human hunting has tended to concentrate on larger edible or economically useful animals as long as they were available—as well as on predators perceived to be a threat to people or livestock—and then to switch to smaller species after the larger ones have been either wiped out or hunted and trapped to the point of extreme scarcity. Nineteenth-century North American market hunters over-hunted a number of species, especially buffalo and elk, ducks, geese and passenger pigeons. By the last decades of that century, the populations of these animals was so reduced that they had become either extinct, uncommon, or so rare as to be placed under legal protection. Among wild birds, once ducks, geese and pigeons had been reduced to the point where it was no longer commercially viable to hunt them, commercial bird hunters started to target smaller, edible species, including shore birds.

Shore birds include creatures like sandpipers, plovers, gulls, terns, auks, skimmers and jaegers. Most members of this group spend a great deal of their time on oceans, lakes, wetlands, beaches or in other open habitats. The largest concentrations of breeding sandpipers, for example, are found in the vast seasonal wetlands and adjacent grasslands of the Arctic tundra in Russia, Canada and Alaska. Among these are a number of species of *curlews*, fairly large sandpipers with long, down-curved bills. At the beginning of the nineteenth century, one species, the Eskimo curlew, was one of the most common shore birds in North America, breeding each year in the Arctic and then undertaking a globe-straddling migration of up to 10,000 miles to productive wintering grounds in South America. Its breeding grounds were uninhabited Arctic grasslands between the tundra and the boreal forest, and it wintered in thinly populated grassland areas in Argentina. Most people were little aware of it except as one among several species of passing migrants in spring or fall, although even as late as 1800 its populations are estimated to have been in the millions.[42]

In the spring, most curlew traveled northwards over the North American Great Plains in large flocks, feeding on grasshoppers and other insects on the tall and mixed grass prairie, and this was where the greatest number fell prey to hunters who would gun them down and ship them back east to market on the railroads. On migration the birds had few natural enemies, so they did not fly away as hunters approached, and their dense flocks meant that high numbers could be killed in a short period of time. After breeding in northwestern Canada and nearby areas, surviving birds flew east across Canada to Labrador and the maritime provinces before starting their long journey to southern South America. In eastern Canada, a different set of hunters was waiting for them. Curlew had been hunted by sportsmen for many years—they were good eating—and their populations had been able to sustain a low level of losses. But once the passenger pigeons began to be wiped out in the 1870s, market hunters turned their attention to other species including the Eskimo curlew, and heavy shooting soon began to dramatically reduce their populations. No one in the late nineteenth century actually recorded the overall number of birds killed by market hunting, but the best estimates are that before their rapid decline, as many as one million curlew were being killed each year. By 1890 they were just about gone, although commercial shipments to Boston continued through the mid-1890s. By 1916, legislation was passed to protect them from further hunting. By then, however, the loss of most of their suitable spring migration habitat through conversion of native grassland to agriculture and the resulting decline in the grasshopper species they used to feed on, made it impossible for their populations to effectively recover. A tiny handful of birds continued to survive throughout the first half of the twentieth century, but records of them grew fewer and fewer. The last confirmed sightings came in 1962 cross-hairs, and by the 1880s, blue whales too began to be intensively hunted in the North Atlantic, where the modest local population was soon nearly wiped out.

By the early years of the twentieth century, the hunt for blue whales had spread around the world, including the cold southern Pacific–Antarctic region. Catches increased, and by the 1920s, it was not only harpooning technology that had begun to change. Industrial whaling now began to feature fleets of fast chase ships, all armed with explosive harpoons, working in conjunction with large factory ships which specialized in processing the catch. The factory ships had large ramps at the stern going down to the water and powerful winches to haul the huge whales on board to be cut up and processed. In 1931 alone, nearly 30,000 blue whales were killed by these fleets, and in the first half of the twentieth century overall, roughly 350,000 of these giants were taken from the sea—fully 99 percent of the entire world population. By the early 1960s, the total estimated global population stood at only 1,000 animals, down from an estimated 300,000– 500,000 before intensive whaling began. By this time, of course, scientific population biology, a specialized field of ecology, had come into existence and environmentalists were able to show convincingly that continuation of the fishery would condemn the blue whale to oblivion. After extensive publicity, and even direct action on the high seas by environmental activists (who interfered with on-going whaling efforts), in 1966 a ban on commercial whaling was finally passed. Since that time, earth's population of blue whales has very slowly started to recover, although even after decades of protection, it still remains at only a small fraction of its former level.[46]

Ocean Fish

In addition to the great whales, the twentieth century also saw significant over-harvesting of many other kinds of marine life. Atlantic cod, for example, was an extraordinarily abundant, commercially valuable cold-water fish that had been the basis of a commercial North Atlantic fishery since the 1500s. But in the 1960s, large *factory bottom trawlers* equipped with electronic fish finders ushered in several decades of extreme over-fishing. In the waters off Newfoundland, for example, catches soared in the 1960s and 1970s, before declining and briefly rebounding in the 1980s, and then completely collapsing in 1992.[47] This led the Canadian Government to close the fishery until 2006 to give the cod population a chance to recover, although when it reopened on a very limited basis that year, cod numbers remained quite low. The likely cause of this persistent decline was that the large commercial bottom trawlers had destroyed so much of the marine ecosystem that they had fundamentally altered the local food web in ways that made it much harder for young cod to survive long enough to breed. Only time will tell whether the Northwest Atlantic cod stocks will ever recover to the point of commercial viability, and some researchers now believe that warming sea temperatures due to increasing atmospheric carbon dioxide will make it very difficult for some depleted cod stocks to recover.[48]

The same decades that saw the over-fishing and collapse of North West Atlantic cod populations also saw significant over-harvesting of a very different kind of fish, the powerful, fast-swimming species of the genus *Thunnus*, known as bluefin tuna. Some of the largest and most powerful of all bony fishes, Atlantic bluefins, for example, can grow to a length of fourteen feet, and weigh up to 1,400 pounds.[49] They live longer than most fish, grow slowly, and take a long time to reach reproductive age. In the second half of the twentieth century and the beginning of the twenty-first, various populations of different bluefin species in the Atlantic, Pacific and Mediterranean were drastically over-fished primarily to supply the demand for Japanese style raw fish. By the last quarter of the twentieth century, bluefin had become a high-priced, globally distributed product with individual large fish sometimes selling for more than $150,000. Before the 1960s, bluefins had experienced only moderate fishing pressure, but starting in the 1970s, growing affluence in Japan, increased consumption of raw fish in other countries, better freezing methods, global air transportation of frozen cargos, increased fishing effort, and the use of enormous long-lines had dramatically increased the catch. Quotas were set by regulatory authorities, but politicians often pressured regulators to set higher quotas than fisheries scientists suggested, and widespread cheating on quotas was also believed to occur in some tuna fisheries. Some fishermen have made money, but bluefin tuna stocks have steadily and precipitously declined, with most down roughly 75 percent. The Mediterranean population is on the verge of collapse with quotas currently much reduced, although some experts believe that they are still too high. Environmental organizations are pushing to have the bluefin listed as an endangered species, but the multi-billion-dollar industry that exploits them is strongly resisting this with lobbying and political pressure.[50]

Conclusion

Hundreds of large or economically valuable species have been over-hunted or over-fished since major Pleistocene population declines began 50,000 years ago. This has disrupted ecosystems on every continent and in every ocean, and numerous species have gone completely extinct. The early twentieth century saw the first stirrings of significant environmental protection especially in some of the richer parts of the world. But it was only around mid-century that the advance of scientific ecology and the rise of a public environmental movement made the conservation of nature something of a global priority. By 1951, for example, an organization called The Nature Conservancy was created in the United States, and since then it has grown to be one of the world's largest private conservation organizations with a mission "to conserve the lands and waters on which all life depends."[51] In 1961 it was joined by the currently even larger World Wide Fund for Nature (WWF), which is based in Switzerland. According to WWF's website, it has offices in more than eighty countries around the world, employs around 6,200 full time staff, is supported by more than five million people and, over the fifty years since it was founded, "has invested around US$11.5 billion in more than 13,000 projects."[52] Around the world today, there are also many hundreds of smaller private conservation organizations, some working very locally, to preserve habitats and species alongside hundreds of governmental agencies at the national, regional and local levels. Some criticisms have been directed at some of the largest private groups concerning their financial management and ties with business, but by and large these organizations play a critically important role in protecting small and large areas of high environmental value and in educating the public about on-going threats facing natural communities. In the future, their work will be essential for the continued survival of species and ecosystems.

International legal agreements have also played a role in helping to preserve species around the world. In 1975, for example, the Convention on International Trade in Endangered Species of Wild Fauna and Flora (CITES; also known as the Washington Convention) came into force, regulating trade in thousands of species that were at risk for serious population decline or extinction.[53] This has certainly made a difference, although it is not unusual for wildlife poaching and smuggling to continue even in countries that have signed on to the convention.

Since the rise of the modern environmental movement, more lands have been set aside from development and some notable species have been protected, although, sadly, rampant over-hunting and over-fishing—some legal and some illegal—continues. Asian tigers, for example, the largest cats to survive the Pleistocene, saw their range dramatically shrink and their wild populations collapse during the twentieth century. In that period, their numbers are estimated to have shrunk more than 95 percent, until only 3,000 tigers remained in the wild by the early twenty-first century, down from hundreds of thousands in the past. Their decline is the result of a combination of past over-hunting for trophies, habitat loss due to rapid human population growth, and uncontrolled (and often illegal) killing of tigers for their skins and other body parts, items widely used in Asian medicine. Saving the tiger has in fact been a major goal of numerous prominent conservation organizations for several decades. Yet the forces arrayed against this great cat are so strong that its populations remain perilously low and continue to decline. Three of its geographical races have already gone extinct in the wild in the twentieth century, one in the vicinity of the Caspian Sea, one on the island of Java and another on the island of Bali in southeast Asia. Several other races are now highly threatened.[54]

The decimation of megafauna and other animals and plants began tens of thousands of years ago. Yet the same short-sightedness and lack of understanding that allowed that to happen is still common today. Many surviving species, large and small, are tottering on the brink of permanent extinction. All too many are likely to go right over the edge.

Notes

1   G. J. Prideaux et al., 2007, An arid-adapted middle Pleistocene vertebrate fauna from south-central Australia, *Nature* 445: 422–425.

2   See Vaclav Smil, 2012, *Harvesting the Biosphere*, MIT Press, and included references.

3   S. P. McPherron et al., 2010, Evidence for stone-tool-assisted consumption of animal tissues before 3.39 million years ago at Dikika, Ethiopia, *Nature* 466: 857–886.

4   M. W. Marzke, 2013, Tool making, hand morphology and fossil hominins, *Philosophical Transactions of the Royal Society B* 368(1630): 20120414.

5   S. Sileshi et al., 2003, 2.6-Million-year-old stone tools and associated bones from OGS-6 and OGS-7, Gona, Afar, Ethiopia, *Journal of Human Evolution* 45(2) (August): 169–177.

6   University of Missouri Museum of Anthropology, Oldowan and Acheulean stone tools, https://anthromuseum.missouri.edu/minigalleries/handaxes/intro.shtml.

7   H. Thieme, 1997, Lower Paleolithic hunting spears from Germany, *Nature* 385: 807–810; M. Balter, 2014, The killing ground, *Science* 344(6188): 1080–1083.

8   D. S. Adler et al., 2014, Early Levallois technology and the Lower to Middle Paleolithic transition in the Southern Caucasus, *Science* 345(6204): 1609–1613.

9   H. Levin, 2006, *The Earth Through Time*, 8th edition, Wiley.

10   J. P. Kempton and D. L. Gross, 1971, Rate of advance of the Woodfordian (Late Wisconsinian) glacial margin in Illinois: stratigraphic and radiocarbon evidence, *Geological Society of America Bulletin* 82: 3245–3250.

11   E. Masters, 2010, Oldest known stone axe found in Arnhem Land, *ABC News*, www.abc.net.au/news/2010-11-05/oldest-known-stone-axe-found-in-arnhem-land/2324852.

12   T. Brown, 1997, Clearances and clearings: deforestation in Mesolithic/Neolithic Britain, *Oxford Journal of Archaeology* 16(2): 133–146.

13   C. Sandom et al., 2014, Global late Quaternary megafauna extinctions linked to humans not climate change, *Proceedings of the Royal Society B* 281: 20133254.

14   T. A. Surovell and B. S. Grund, 2012, The associational critique of Quaternary overkill and why it is largely irrelevant to the extinction debate, *American Antiquity* 77(4): 672–687.

15   M. Rasmussen et al., 2011, An Aboriginal Australian genome reveals separate human dispersals into Asia, *Science* 334(6052): 94–98; A. Beyin, 2011, Upper Pleistocene human dispersals out of Africa: a review of the current state of the debate, *International Journal of Evolutionary Biology* 2011: 615094.

16   G. J. Prideaux et al., 2007, An arid-adapted middle Pleistocene vertebrate fauna from south-central Australia, *Nature* 445: 422-425.

17   C. Sandom et al., 2014, Global late Quaternary megafauna extinctions linked to humans not climate change, *Proceedings of the Royal Society B* 281: 20133254.

18   S. Rule et al., 2012, The aftermath of Megafaunal extinction: ecosystem transformation in Pleistocene Australia, *Science* 335(6075): 1483–1486.

19   Seaworld, Elephants, diet & eating habits, https://seaworld.org/en/animal-info/animal-infobooks/elephants/diet-and-eating-habits.

20   C. Sandom et al., Global late Quaternary megafauna extinctions linked to humans, not climate change, *Proceedings of the Royal Society B* 281: 20133254.

21   Ibid.

22   R. D. E. MacPhee (ed.), 1999, *Extinctions in Near-Time: Causes, Contexts and Consequences*, Springer.

23   M. E. Allentoft et al., 2014, Extinct New Zealand megafauna were not in decline before human colonization, *PNAS* 111(13): 4922–4927.

24   Animal Diversity Web, *Hydrodamalis gigas*, http://animaldiversity.org/accounts/Hydrodamalis_gigas.

25   IUCN Red List of Threatened Species, *Pinguinis impennis*, www.iucnredlist.org/details/full/22694856/0.

26   Ibid.

27   Worldwatch Institute, 2013, Mammal species decline in face of over-hunting, habitat loss, climate change, www.worldwatch.org/mammal-species-decline-face-over-

hunting-habitat-loss-climate-change; PBS, The National Parks: Everglades National Park, www.pbs.org/nationalparks/parks/everglades.

28 R. Wakeman, 2014, Twenty gauge versus twelve gauge: what is best?, www.chuckhawks.com/twenty_vs_twelve.htm.

29 Wikipedia, Punt gun, https://en.wikipedia.org/wiki/Punt_gun; Rare Historical Photos, Punt Gun, http://rarehistoricalphotos.com/punt-gun-used-duck-hunting-banned-depleted-stocks-wild-fowl.

30 US History, Early American Railroads, www.ushistory.org/us/25b.asp.

31 Smithsonian Institution, History of the telegraph, http://historywired.si.edu/detail.cfm?ID=324.

32 PBS—History Detectives, 2015, Gun Timeline, www.pbs.org/opb/historydetectives/technique/gun-timeline.

33 Smithsonian, The passenger pigeon, www.si.edu/encyclopedia_Si/nmnh/passpig.htm.

34 Wild Birds, Extinction of the American passenger pigeon, www.wildbirds.org/apidesay.htm.

35 W. T. Hornaday, 1889, The extermination of the American Bison, *Project Gutenberg*, www.gutenberg.org/files/17748/17748-h/17748-h.htm#ii_iii_b_3.

36 US National Park Service, Tallgrass Prairie National Preserve, 2015, Nature, www.nps.gov/tapr/learn/nature/index.htm; H. W. Polley et al., 2005, Patterns of plant species diversity in remnant and restored tallgrass prairie, *Restoration Ecology* 13(3): 480–487.

37 A. K. Knapp et al., 1999, The keystone role of bison in North American tallgrass prairie, *Bioscience* 49(1): 39–50.

38 US National Archives, The Homestead Act of 1862, www.archives.gov/education/lessons/homestead-act.

39 US Library of Congress, Pacific Railroad Act, https://en.wikipedia.org/wiki/Pacific_Railroad_Acts.

40 Historynet.com, Westward expansion, www.historynet.com/westward-expansion.

41 W. T. Hornaday, 1889, The extermination of the American Bison, *Project Gutenberg*, www.gutenberg.org/files/17748/17748-h/17748-h.htm#ii_iii_b_3; The Nature Conservancy, 2014, South Dakota studying bison DNA, www.nature.org/ourinitiatives/regions/northamerica/unitedstates/southdakota/howwework/studying-bison-dna.xml.

42 Alaska Department of Fish and Game, 2014, Eskimo curlew, www.adfg.alaska.gov/index.cfm?adfg=eskimocurlew.main.

43 Bird Life International, 2014, Eskimo curlew, www.birdlife.org/datazone/species-factsheet.php?id=3008.

44 J. R. McNeill, 1994, Of rats and men: a synoptic environmental history of the island Pacific, *Journal of World History* 5(2) (Fall): 299–349.

45 Wikipedia, Svend Foyn, https://en.wikipedia.org/wiki/Svend_Foyn.

46 American Cetacean Society, Blue whale, http://acsonline.org/fact-sheets/blue-whale-2; IUCN Red List of Threatened Species, *Balaenoptera musculus*, www.iucnredlist.org/details/full/2477/0.

47 MARINEBIO, Atlantic cod, *Gadus morhua*, http://marinebio.org/species.asp?id=206.

48 US National Oceanographic and Atmospheric Administration, 2014, Atlantic cod, www.nefsc.noaa.gov/sos/spsyn/pg/cod; A. J. Pershing et al., 2015, Slow adaptation in the face of rapid warming leads to collapse of the Gulf of Maine cod fishery, *Science* 350(6262): 809–812.

49 IUCN Red List of Threatened Species, *Thunnus thynnus*, www.iucnredlist.org/details/full/21860/0.

50 Ibid.

51 The Nature Conservancy, About us: vision and mission, www.nature.org/about-us/vision-mission/index.htm?intc=nature.fnav.

52 Worldwide Fund for Nature, Quick facts, http://wwf.panda.org/wwf_quick_facts.cfm.

53 Convention on International Trade in Endangered Species of Wild Fauna and Flora, https://cites.org.
54 IUCN Red List of Threatened Species, *Panthera tigris*, www.iucnredlist.org/details/full/15955/0.

2

OVER-HARVESTING OF NATURAL RESOURCES

Fresh Water

Introduction

When life first appeared more than three billion years ago, it did so in the presence of water. Only in the last half billion years or so have living things become well adapted to living on dry land. For creatures making the difficult transition out of the ocean or fresh water, one of the greatest challenges lies in obtaining and retaining adequate water in their tissues. Among the vertebrates, the earliest forms which spent significant time on land were the ancestors of today's amphibians, most of which continue to live in wet or damp habitats. The first vertebrates that were truly successful in dry environments were the reptiles, and they managed this by developing a range of improved water-conservation structures, especially water-proof skin and eggs, internal lungs, and water-conserving kidneys. When birds and mammals later arose from reptiles, those water-conserving structures were retained, with mammals eventually dispensing even with waterproof eggs by keeping the babies inside the mother until they *hatched*.

Over hundreds of millions of years of life on land, different creatures adapted to the varying amounts of water available in the environments where they lived, as well as to the low levels of salt present in most water available on the land. In perennially wet habitats like swamps and marshes, advanced water conservation was scarcely necessary, while in the driest deserts, highly efficient water retention often made the difference between life and death. The majority of creatures evolved in moderately damp habitats, where some degree of water conservation was very helpful but the highly specialized adaptations needed for successful desert life were not needed. A few desert vertebrates like the North American kangaroo rat evolved such extreme drought tolerance that they never need to drink at all, and instead can derive all of their water metabolically from their food. But human ancestors did not evolve in dry deserts, and so they never developed such highly specialized water-conservation mechanisms. As a result, in order to remain healthy, every human needs to take in fresh water in the form of food and drink every day, generally two to three liters.[1] And for plants, no matter how well adapted they become to drought or desert conditions, they still require some water because without it they cannot make the carbohydrates which are both their food and the raw materials with which they build their bodies.

Farming and Water

Before our ancestors began to intentionally plant food crops, our hunter-gatherer forbears used only small amounts of water, primarily for drinking, cooking, processing food, and tool-making. They often camped near a stream, lake, river or spring, and given their low population density, getting enough safe drinking water was probably not too much of a problem. Starting around 10,000 years ago, however, agriculture led to significant increases in human populations and began to tie people to settlements; in drier climates, it also created a need for large quantities of water to irrigate fields, most of which was initially obtained by the diversion of river water (see Chapter 3). This was the case in numerous ancient societies including Mesopotamia, Egypt, and parts of China and India. Water that was already present on the surface was the easiest to tap, but during prolonged droughts, rivers and ponds often dried up, and people soon found that when this happened, they could still find some water by digging holes in dry riverbeds. This discovery—more than 8,000 years ago—led to the first intentionally dug water wells, giving people access to supplies of water present beneath the land's surface, where it was less subject to evaporation. It also led to a basic distinction between surface water, such as that found in a river or lake, and groundwater, which is usually not visible on the surface, but can be accessed by digging or drilling.[2]

Sources of surface water—lakes, ponds, rivers, streams, marshes and man-made reservoirs—are the easiest to exploit and are often the first to be utilized, but they are limited in volume and run a comparatively high risk of pollution. Underground water, on the other hand, is about one hundred times as abundant as surface water. Most of it is found in porous sediments or in cracks within masses of rock. In places, it even bubbles to the surface in springs, but frequently it can only be accessed by digging or drilling wells. Even once wells have been dug, however, constant work needs to be done to lift groundwater to the surface, and sometimes the water itself is salty or *hard*, and contains large amounts of dissolved calcium, magnesium and other ions which reduce its suitability for some domestic and industrial uses.[3]

After the start of farming 10,000 years ago, human population was able to gradually expand, and as it did it displaced more and more pre-existing plant communities with ever larger areas being converted to cultivated fields and grazing land. During farming's first 8,000 years, human population increased about fifty times—from roughly 5 million to 250 million.[4] Over this same period, agricultural fields and pastures must have also expanded enormously. In areas where irrigation made farming more productive, larger and larger quantities of water were diverted and applied to fields. In some hot dry climates, like Mesopotamia, extensive use of irrigation employing reasonably hard water, also led to mineral salt build-up in the soil which slowly reduced fertility.[5]

Irrigation in dry climates may also cause salt accumulation if so much fresh water is applied to fields that it percolates down to contact shallow, underlying aquifers. Some of these may contain saline water, and if the irrigation-raised water table rises into the plant root zone, salt can move upward with the water table and becomes deposited near the surface. Avoiding these problems under long-term irrigation is difficult. To do so, large amounts of high quality water must be continually applied, mostly to nourish crops, but some also to simply to flush out the salts accumulating from earlier irrigation. Salt-laden drainage must then be diverted away from both the fields and from the well areas providing the irrigation water. Under such conditions, additional fertilizer must also continually be applied because the large amount of irrigation water is constantly leaching available nutrients out of the soil.[6]

Agriculture was such a powerful, overwhelming set of practices in relation to the wild plants and animals being displaced, that human numbers continued to grow and natural habitats continued to be destroyed even after agriculture's first 8,000 years. Less than 2,000 years after they hit 250 million in the Year 1, human population grew to be four times as large in 1800, when roughly one billion people were present (see Chapter 3). Even though much of the expansion of farmland over this long period occurred in temperate zone areas where crop fields generally were not irrigated, water use still increased very fast, not only to meet the requirements of additional people, but also the needs of the vastly increased populations of horses, oxen and other farm animals.

Until relatively recently, simple farm irrigation and domestic use were the main consumers of water. Little by little, however, other uses were found, with cities of the Indus River Valley Civilization pioneering the use of flowing water to flush out sewage.[7] Roman civilization dramatically expanded all these activities, diverting huge quantities of water for mining, bathing and carrying away human wastes, and it did so with little concern for the ecology of the source streams or the damage the dumped wastes might cause to natural aquatic systems. Roman engineers, in fact, often diverted entire streams to create a powerful erosive force for mining which they used both to wash away overlying sand and gravel and to separate particles of metal ore from surrounding materials. In Europe, they did this on a particularly large scale in Spain, but also in Wales and other places.[8]

Some progress was made in ancient times in our understanding of pumps as shown by the work of inventors like Ctesibius (285–222 BC) and Archimedes (287–212 BC). Archimedes' design, featuring a screw turning within a cylinder, came to be widely employed. Ancient Rome successfully channeled many mountain springs and streams through stone-lined aqueducts to provide water for their cities, constructing roughly 600 aqueducts over five or six centuries.[9] Some of these remained in use after the collapse of the Roman Empire, but in most places water to meet human needs was later obtained more locally and in a much more basic way through the use of simple mechanical devices like water wheels, buckets on a rope attached to a windlass, or augers turning inside cylinders. Most of the water used was obtained directly from surface springs, streams or lakes, or else from shallow hand-dug wells. The power to raise and move water around was mostly supplied by muscle power—human or animal.

Invention of Modern Water Pumps

Modern science is usually thought to have originated in the sixteenth and seventeenth centuries, an era when the lines separating pure science and technology were not as sharp as they are today. As they experimented with gases and liquids inside vessels, many seventeenth-century researchers simultaneously tried to develop improved pumps, with the Englishman Samuel Moreland, for example, patenting an early plunger type in 1675.[10] Around 1690, the French scientist Denis Papin outlined the theory of a highly efficient centrifugal pump, a device which could move water by having a pin-wheel like structure spinning inside a closed casing, and by the 1750s, the Swiss mathematician/physicist Leonhard Euler had developed an accurate mathematical model of such devices. Euler's life (1707–1783) spanned the period when the first successful steam engines were introduced to pump water out of mines, and also when early steam engines began to be improved, most notably through the work of James Watt and Matthew Boulton between 1765 and 1790 (see Chapter 6). Workable steam engines made it easier not only to keep underground mines dry enough to work in, but also to provide safer drinking water through municipal water supply systems which typically pumped river water up to higher reservoirs from where it could flow through pipes by gravity to consumers.[11]

Around the same time, the industrial revolution also led to the opening of large numbers of new industrial facilities, many of which required large amounts of water both to power machinery and to process textiles, chemicals and other materials. In many cases these new factories withdrew large amounts of clean water from rivers and other water bodies, some of which was lost in the manufacturing process or shipped away in finished products. Most, however, was discharged right back into the same ponds and streams from which it had been taken, often loaded down with all the waste products and toxic chemicals used in the manufacturing process. In many cases this caused significant damage to both aquatic ecosystems and the land areas adjacent to them (see Chapter 4).

The nineteenth century saw dramatic progress not only in the design of pumps, but also in the power plants needed to run them. In the first half of the century improvements in design, metallurgy and manufacturing led to better pumps—one demonstrated at the Crystal Palace Exhibition in London in 1851, for example, was 68 percent efficient. During most of that century, reciprocating steam engines were the prime movers driving most water pumps, but by the century's second half the newly developed internal combustion engines had started to replace them, a trend which continued in the twentieth century along with the introduction of electrically driven pumps (see Chapter 8 and Chapter 6 for related developments in the design of water turbines and internal combustion engines). By the late nineteenth century it became possible to create very powerful water pumps that could be run independently of any of the earlier common sources of power—wind power, running water, animal power or human muscles. It had also become feasible, if not always easy or cheap, to raise up and move around very large quantities of water.

The 1800s saw the addition of three quarters of a billion more people as land clearing, agriculture and food supplies expanded (especially in the Americas and Australia), and safer drinking water supplies and better understanding of infectious disease reduced mortality. By 1900, the amount of water needed to support human society, including farm and industrial usage, had grown so large that rather than calculate it in liters or gallons, it became simpler to estimate it in terms of cubic kilometers, an astonishingly large unit equal in volume to one thousand billion liters (a liter is a bit more than a quart). One cubic kilometer is enough water to fill a canal one meter wide, one meter deep and more than 600 miles (1000 kilometers) long.

Estimates of Water Use

At the start of the twentieth century, worldwide freshwater use had already risen to about 500 cubic kilometers annually, even though the greatest growth in water withdrawals was still to come. As the century wore on, rural electrification in developed nations made it possible for powerful electric water pumps to be installed in many new locations, while the development of dependable diesel engines allowed pumping to be done in more remote areas away from power lines. These new sources of power combined with improved pumps to allow significantly increased quantities of water to be withdrawn from ground and surface supplies for irrigation, industry and domestic use. By the middle of the twentieth century, it had become possible to pump groundwater out of aquifers (underground water sources) much faster than nature could refill them. And new drilling techniques, some borrowed from the petroleum and natural gas industries, also provided access to previously unavailable deep aquifers, often hundreds of feet down, some containing *fossil water* that had accumulated thousands of years earlier during periods when local climates were significantly wetter. In developed nations, even recreation began to demand significant amounts of water in some locations as numerous new golf courses were designed with regular irrigation, and a large number of downhill ski resorts began pumping large quantities of water up to their slopes to create artificial snow. (Some of the water used for these purposes eventually ended up replenishing local groundwater and streams, but a great deal of it also disappeared through evaporation.)[12]

Over the course of the twentieth century, demand for water steadily increased—despite a lull during the economic depression of the 1930s—with most additional supplies going to meet the needs of farm irrigation, power and industrial plants, mushrooming human populations and increasing standards of living. In this period, global population more than tripled from one and three quarter billion in 1900 to about six billion in 2000. In the same time frame, use of water in agriculture worldwide grew five-fold, while total water consumption increased even faster. By 2010, worldwide human water consumption had increased more than 800 percent over the amount used in 1900 to about 4,000 cubic kilometers per year. This is an enormous amount of water, enough to completely refill the vast basin of North America's Lake Superior—the third largest lake by volume in the world—in just three years.[13]

It is easy to view all this as progress, as it is hard to be against better human survival or greater food production, although in the long run some water-intensive industrial activities, especially in the chemical and paper industries, probably represent a net loss for both humanity *and* nature. In fact, though, people's quest for water ended up going far beyond what nature could support, and in many places water sources actually began to dry up, while natural communities dependent on water that was being diverted and withdrawn suffered damage, decline and in some cases even complete collapse.

In wetter areas of the world, most water usage goes for domestic and industrial uses. In such climates, industrial facilities—especially power plants—often discharge much of the water they obtain, although their releases are frequently contaminated with chemical and biological pollutants, or—in the case of electricity generation—with heat. In drier climates, on the other hand, there is frequently not enough rain to support large human populations or to grow abundant crops, and in the twentieth century massive water transfer and irrigation systems were undertaken in many countries often with little thought to their longer-term environmental consequences. The impacts of these schemes have varied with the sources of the diverted waters and with the final uses to which they were put. By 2000, an estimated 279 million hectares (1,069,503 square miles) of land was under artificial irrigation worldwide, an area nearly as large as modern-day India.[14] Over this vast irrigated area, roughly 60 percent of the water used was withdrawn from rivers and other surface water bodies, while nearly all of the remainder was pumped up from underground.[15]

The most striking impact of such massive water diversions was the drying out of many important surface river drainages, especially in more arid or heavily populated parts of the world. In the long run, though, the damage caused by the over-pumping of underground waters may prove to be just as great. Such extreme withdrawals of water impact nature in different ways, and the most immediate effect is usually a decline in the community of living things inhabiting the river, lake, or other body of water. Highly productive wetlands, for example—marshes, swamps and shallow backwaters—often simply disappear when groundwater is over-pumped; estuaries, places where river waters used to mix with ocean water to create rich habitats that were ideal for the breeding of fish, invertebrates and water birds, often become more salty and no longer able to play their critical environmental roles.[16]

In Asia, North America, Australia and Africa, several large and important rivers have been profoundly damaged, and sometimes even completely drained dry. The Indus in Pakistan is one, and the Colorado and the Rio Grande/Rio Bravo in North America are two others. All three of these sometimes now run dry now as they approach the sea. The Nile in Africa, China's Yellow River, India's Ganges and the Murray-Darling in Australia have all been heavily degraded by extensive water withdrawals even though there is usually still some flow out of their mouths.[17] And in another area of Asia, excessive freshwater diversions from drainages which used to flow into the brackish Aral Sea have caused one of the worst environmental catastrophes of modern times.

The Aral Sea Disaster

In the middle of the twentieth century, the landlocked, slightly salty Aral Sea was the fourth largest lake in the world in area. It was located in a dry region of Central Asia, in Uzbekistan and Kazakhstan, and was maintained by fresh waters flowing into it from several rivers that drained large surrounding mountainous areas. In the 1950s, it covered an area of more than 26,000 square miles, an area roughly equivalent to the Irish Republic; it was environmentally rich and supported a thriving fishing industry. Over the following decades, however, the central planners of the former Soviet Union ordered the construction of several large dams on the rivers feeding it, along with an extensive series of poorly built, open irrigation canals. Their aim was to divert almost all of the water from the Aral's two main sources of fresh water, the Amu Darya and Syr Darya Rivers, onto to dry semi-desert land in order to grow water intensive crops such as rice and cotton. The Soviet planners seem to have known that their schemes would cause the Aral Sea to dry up, but they went ahead anyway.[18]

Under the impact of these diversions, the Aral's area and its volume of water both decreased by roughly 90 percent between 1960 and the late 1990s. In that same period, its salinity increased up to twenty times as evaporation concentrated its dissolved salts.[19] The productive and economically important fishery it had supported in the first half of the twentieth century disappeared, along with the livelihood of the fishermen and their families, and many unique native animals and plants. In place of the Aral Sea, all that remained were a few separate, much smaller water bodies located in the basins where the deepest parts of the original lake had been. In the early twentieth century, the Aral had been only one third as salty as the ocean, but now the shrinking waters in the surviving lake fragments became up to six times as salty as seawater, too salty to support any aquatic life except microbes. Twenty of the twenty-four native fish species vanished, and the salinity eventually became so extreme that even highly salt-tolerant invertebrates were no longer able to live in them.

Lacking its freshwater inflows, most of the old lake bed became dry land, but it was land that could not be used for productive purposes. The old lake bed revealed by the water's retreat was thickly encrusted with salt and toxic chemicals that had been present in the mud on the lake bottom. Salt had condensed out of the shrinking lake waters themselves, while the chemicals had come from a combination of agricultural runoff (fertilizers and pesticides), industrial wastes and weapons-testing.[20] In most areas on the exposed lake bed, not even desert vegetation would grow, and drought and winds frequently created damaging dust storms over a wide region, spreading toxic substances which caused widespread health problems in several poor countries in Central Asia. Some of the dust even reached high mountain glaciers in the Himalayas, darkening their surface and further accelerating melting that was already increasing due to global warming. Residents of the region have seen severe health effects: infant mortality rates have roughly tripled; developmental problems and low birth weight have both become more common; numerous kinds of disease have become more prevalent including respiratory and gastrointestinal conditions, diseases of the kidneys and liver, and even cancer.[21]

Over-Pumping of Groundwater

The disaster in the Aral basin was caused by diversion of surface waters, but the twentieth century also saw the beginning of extraordinary exploitation of groundwater as petroleum extraction techniques, internal combustion engines and electric motors made deep drilling and long-distance pumping feasible. Geological formations in which wells are able to provide access to significant amounts of groundwater are known as *aquifers*, literally *water bearers*, and they are present in many areas of the earth where soils and rocks have allowed downward infiltration of precipitation. Most aquifers are nothing like underground rivers or streams; rather, they resemble thick, rigid landscape-wide sponges, in which the rocks, sands and gravels beneath the earth's surface make up the body of the sponge, with the water filling most of the space between the solid particles. The largest individual aquifers extend for hundreds of miles, and some are many hundreds or even thousands of feet deep. In many areas, aquifers are continually being refilled by infiltration of new precipitation; in other areas, however, where the climate has become drier over time, they often store water that accumulated during earlier, wetter times—so-called *fossil water*. Exploiting this is more like mining than obtaining normal water supplies; like mineral ore, once pumped out, such non-renewable sources of water are simply gone. In ultra-dry Saudi Arabia, for example, in recent decades wheat was grown by irrigating it with fossil water drawn up from depths as great as 4,000 feet, although after a number of years of this very expensive effort, the large government subsidies that had encouraged it were withdrawn.[22]

If groundwater is removed at the rate at which it is naturally replenished, it can be an excellent source of water to meet human needs. It is often of high quality due to natural filtering through the minerals in the earth (although at times it may also contain an excess of soluble minerals like calcium, and so be known as *hard water*, which may have to be *softened* before it is suitable for pipes, plumbing and some uses). It is generally free of surface contaminants like dust and debris, and its moderate removal from the ground does not damage the ecology of local surface waters like rivers and lakes.

If, on the other hand, it is *over-pumped*, or removed faster than it is being replenished, a whole range of problems can result. First, the water itself is gradually used up as its level in the aquifer falls, requiring wells to be deepened and much more energy to be used to raise the water to the surface. Second, the quality of the water often falls. This can happen either as the result of increased surface contamination when runoff carrying agricultural chemicals mixes in a smaller volume of groundwater, or else from greater mixing with the deepest groundwater layers, which often contain significant amounts of salt (fresh water generally floats on top of salt water). Third, natural water levels in rivers, lakes and streams may decline as the drying out of mineral layers underlying them causes their waters to percolate downwards, and this can heavily damage the living communities present in the surface waters and in the rich riverbank and lowland communities adjacent to them. Fourth, the land surface itself may begin to sink as the weight of the layers of sand and gravel causes tighter packing together of particles once the supporting layers of water have been removed, a process known as subsidence. In urban areas like Mexico City, this frequently damages buildings, or even causes sewers to stop working because gravity sewers will not flow uphill. Shanghai, China is another large city that has experienced sinking land from over-pumping, as have many heavily irrigated agricultural areas including California's San Joaquin Valley. Fifth, over-pumping of fresh water, especially in coastal areas, often leads to horizontal mixing in of very salty ocean water, which can sometimes make the water from the original fresh water wells undrinkable. And finally, significant amounts of groundwater that are pumped up for irrigation eventually end up in the world's oceans, where they make a contribution to continually rising sea levels that threaten coastal areas.[23]

Fossil water has been heavily exploited in many parts of the world since the middle of the twentieth century. In parts of Saudi Arabia, Libya, Northern China, India, Australia and the United States, well water has been drawn from aquifers that have stored it for thousands of years, since periods when local climates were significantly wetter. In such areas today, groundwater is either not being replenished by precipitation at all, or is being replenished only at a low rate. On the North China Plain, for example, over-pumping of groundwater caused the water table to fall from 12 meters deep to 32 meters in the last decades of the twentieth century, while the annual surface flow in the nearby Fuyang River declined roughly 80 percent over the same period.[24] In many other places, heavy pumping is causing water tables to fall as rapidly as several feet per year. In India this has even led to a mini-epidemic of suicides among poorer farmers who have lost their livelihoods as their shallow drilled or hand-dug wells have run dry after nearby deeper wells artificially lowered the local water table.[25]

As this is being written, worldwide human population is roughly seven billion, and is expected to rise to nine billion by the middle of the twenty-first century. In the next few decades, food production will have to increase significantly, not only to feed two billion additional people—a number greater than all of humanity in 1900—but also to accommodate the increasing demand for animal protein which generally accompanies rising standards of living. Meeting this increased demand for food will further strain limited freshwater supplies. According to the British Institute of Mechanical Engineering, growing enough grain to make a kilogram of pasta or bread requires an average of about 1,700 liters of water; for a kilo of rice, it's 2,500 liters. Producing a kilogram of chicken meat, on the other hand, uses up 4,300 liters of water; of pork, 6,000 liters; and of beef, 15,400 liters.[26] Beef cattle generally require between 20 and 55 liters (5–15 gallons) of drinking water each day.[27] With increasing populations in many countries, and growing demand for more water-intensive foods, it is easy to see that water-scarce parts of the world will see shortages increasing as time goes on.

By the early twenty-first century, nearly all of earth's best farmland had already been wrested from nature, and in many regions water has become the major limiting factor in agricultural production. Worldwide, crop irrigation uses up 70 percent of all water supplies and leads to the production of 40 percent of all our food. At least 20 percent of irrigation water currently comes from non-sustainable supplies, and environmental damage is increasing as more and more extreme measures are taken to procure water for crops.[28] In the Near East, South Asia and Northeast Africa, several regions are already experiencing political tensions over the allocation of scarce water resources, as is occurring between Egypt and Ethiopia, and there is a distinct possibility that in coming decades wars will be fought over water resources, especially if climate change leads to increased droughts in heavily populated parts of the world.

Conclusion

Our current utilization of water resources is in desperate need of improvement. In many parts of the world a large fraction—in some places even the majority—of irrigation water is *wasted*, lost to leaks in distribution and to excessive evaporation due to inefficient application. The efficiency with which water is used can be enhanced many times over through training for farmers, better infrastructure and different water application methods. Governments, in fact, have often encouraged water waste by using public monies to build water infrastructure and then providing water to farmers either for free or at far below its true cost, giving them little incentive to use it more efficiently. Less water-intensive crops could profitably be grown in many places where irrigation is now extensively used, and plant breeding could produce new varieties that require less water in the first place. People in richer countries could also help by eating more grains, vegetables and fruits, and less meat—it would probably lead to better health as well. Doing this would save large amounts of water by reducing the huge amounts of grain currently consumed by meat animals.

World population is enormous and still increasing, and the earth simply does not have enough affordable, high quality fresh water in the right locations to satisfy all human needs. In the Aral Sea, the Indus, the Colorado, the Yellow and the Murray–Darling, our thirst for water has already heavily damaged natural communities, reducing their ability to support human civilization in the future. In coming decades, the measures people are likely to take to meet their ever-increasing demand for water will almost certainly cause significant additional harm to the environment. To counter this, very strenuous efforts need to be made to achieve efficiency, innovation and moderation in the ways that water is used.

Notes

1   Mayo Clinic, Water, www.mayoclinic.org/healthy-lifestyle/nutrition-and-healthy-eating/in-depth/water/art-20044256.
2   J. van der Gun, 2012, Groundwater and global change, www.zaragoza.es/contenidos/medioambiente/onu/789-eng-ed4-res2.pdf.
3   US Geological Survey, The world's water, http://water.usgs.gov/edu/earthwherewater.html; World Health Organization, Hardness in water, www.who.int/water_sanitation_health/dwq/chemicals/en/hardness.pdf.
4   Wikipedia, World population estimates, https://en.wikipedia.org/wiki/World_population_estimates.
5   http://mygeologypage.ucdavis.edu/cowen//~GEL115/115CH17oldirrigation.html; T. Jacobsen and R. M. Adams, 1958, Salt and silt in ancient Mesopotamian agriculture, *Science* 128: 1251–1258.
6   R. Cowen, University of California, Davis, My geology page: ancient irrigation, http://mygeologypage.ucdavis.edu/cowen//~GEL115/115CH17oldirrigation.html.
7   U. Singh, 2008, *A History of Ancient and Early Medieval India: From the Stone Age to the 12th Century*, Pearson Education, pp. 151–155.
8   S. Timberlake, 2004, Early leats and hushing remains suggestions and disputes of Roman mining and prospection for lead, *Mining History: The Bulletin of the Peak District Mines Historical Society* 15(4/5): 64–76, www.researchgate.net/publication/279695917_Early_leats_and_hushing_remains_suggestions_and_disputes_of_Roman_mining_and_prospection_for_lead.
9   W. D. Schram, 2014, Literature on Roman aqueducts, www.romanaqueducts.info/aqualib/aqualit.htm.
10 M. Segrest, The history of pumps: through the years, www.pump-zone.com/topics/pumps/pumps/history-pumps-through-years.
11 Ibid.; Wikipedia, History of water supply and sanitation, https://en.wikipedia.org/wiki/History_of_water_supply_and_sanitation.
12 University of Michigan, 2006, Human appropriation of the world's freshwater supply, www.globalchange.umich.edu/globalchange2/current/lectures/freshwater_supply/freshwater.html; see also World Resources SIM Center, 2012, Global water consumption 1900–2025, www.wrsc.org/attach_image/global-water-consumption-1900-2025.
13 Wikipedia, Lake Superior, https://en.wikipedia.org/wiki/Lake_Superior.
14 S. Siebert et al., University of Bonn, 2006, The digital global map of irrigation areas, www.tropentag.de/2006/abstracts/full/211.pdf.
15 S. Siebert et al., University of Bonn, 2013, Upgrade of the digital global map of irrigation areas, www.lap.uni-bonn.de/research/downloads/gmia/siebert_et_al_2013_gmia5; Wikipedia, Irrigation, https://en.wikipedia.org/wiki/Irrigation.
16 H. R. Johnson and R. H. Webb, 2013, The great Mesquite forest: a permanent loss in avifauna along the Santa Cruz River in the Tucson Basin, *Vermillion Flycatcher* 58(2): 18–19.
17 National Geographic, 8 mighty rivers run dry from overuse, http://environment.nationalgeographic.com/environment/photos/rivers-run-dry.

18 Columbia University, The Aral Sea crisis, www.columbia.edu/~tmt2120/introduction.htm.

19 P. Micklin, 2010, The past, present and future Aral Sea, *Lakes and Reservoirs: Research and Management* 15(3): 193–213; for a time lapse of the lake's disappearance, see Wikipedia, Aral Sea, https://en.wikipedia.org/wiki/Aral_Sea.

20 Wikipedia, Aral Sea, https://en.wikipedia.org/wiki/Aral_Sea.

21 P. Whish-Wilson, 2002, The Aral Sea environmental health crisis, *Journal of Rural and Remote Environmental Health* 1(2): 29–34, www.jcu.edu.au/jrtph/vol/v01whish.pdf.

22 L. Brown, 2008, As countries over-pump aquifers, falling water levels mean falling harvests, June 18, www.treehugger.com/files/2008/06/countries-aquifers-water-shortage.php.

23 University of Miami, Rosenstiel School of Marine and Atmospheric Sciences, press release, May 2013, Extensive land subsidence in Central Mexico caused by groundwater extraction, http://moa.agu.org/2013/files/2013/05/ChaussardWdowinski_Mexico AGU_FINAL1.pdf; Wikipedia, Groundwater-related subsidence, https://en.wikipedia.org/wiki/Groundwater-related_subsidence.

24 L. F. Konikow and E. Kendy, 2005, Groundwater depletion: a global problem, *Hydrogeology Journal* 13: 317–320.

25 L. Brown, Earth Policy Institute, 2013, Peak water: what happens when the wells go dry?, www.earth-policy.org/plan_b_updates/2013/update115; A. Waldman, 2004, Debts and drought drive India's farmers to despair, *New York Times* (June 6), www.nytimes.com/2004/06/06/world/debts-and-drought-drive-india-s-farmers-to-despair.html.

26 Datablog, 2014, How much water is needed to produce food and how much do we waste?, *The Guardian* (January 10), www.theguardian.com/news/datablog/2013/jan/10/how-much-water-food-production-waste.

27 Clemson University Cooperative Extension, 2014, Food and water requirements for livestock, www.clemson.edu/extension/ep/food_water_req.html.

28 Y. Wada, 2012, Non-sustainable groundwater sustaining irrigation, *Global Water Forum* (February 13), www.globalwaterforum.org/2012/02/13/non-sustainable-groundwater-sustaining-irrigation.

3

PEOPLE AND THE LAND

Introduction

Of all common human activities, the occupation and utilization of land in response to growing population has probably had the greatest impact on earth's natural living communities. Even compared to other damaging technologies, the conversion of more and more of the planet's surface to farming, grazing and plantations has almost certainly led to the largest losses of plant and animal life. Land clearing and agriculture are the most obvious destructive activities, but the rampant geographical spread of people around the world, driven by colonization, migration and the development of global travel and trade has also had significant impacts. In addition to the outright destruction of many pre-existing living communities for agriculture, the movement of people and goods has also relocated thousands of different species from home ecosystems in which their growth and spread was well controlled by evolved checks and balances, into new, far-flung locations where their unrestrained reproduction often leads to insoluble problems of species invasion. The complex interactions between human numbers, global expansion, and land use are challenging to unravel. But they are critically important if we want to understand the environmental impacts of human life and the possibilities for eventually coming into balance with the earth.

Food Supply and Population

It is hard today to imagine a time when people did not intentionally plant seeds and then harvest a crop at the end of a growing season. Yet, as familiar as this cycle is to us, it is still a relatively recent development in human life. Before about 10,000 years ago, there is little or no evidence of intentional growing of food. Instead, our ancestors lived as hunter–gatherers, killing and eating animals by hunting, fishing and trapping while also gathering as much plant food as possible—fruits, berries, seeds, roots, stems or leaves—within range of their mobile camps.[1] In a limited way, some of their food could be preserved for later use by smoking or drying, and in cold climates some could be frozen over the winter. But most food was probably consumed fresh, within a few days or weeks of acquiring it. People likely ate a great deal when large quantities were available—perhaps temporarily gaining weight—and then survived on a great deal less when food was scarce. In a few places around the world, small groups of people continued to live like this into the nineteenth and twentieth centuries, making it possible for the emerging discipline of anthropology to document their way of life.[2]

Before the start of farming roughly 10,000 years ago, the best estimate is that human numbers worldwide amounted to less than ten million. By the year 2000, however, the number had grown to exceed 6,000 million (six billion) with several billion more expected to be added over the course of this century. Ten thousand years may seem like a large span of time, but in earth's long history, it was the mere blink of an eye. This enormous expansion of human population represents a revolutionary change in the living world, a development that affects almost everything else. For thousands of years after the start of farming, populations grew relatively slowly in response to gradually increasing food supplies as more natural habitat was converted and farming technologies gradually advanced. Overall however, this was a process of two steps forward and one step back, with periodic losses due to disease, famine, war, weather and declines in soil fertility. From the 1700s on, however—and especially after 1800—human numbers began to grow much more rapidly as scientific knowledge started to bring down human death rates while also revolutionizing agriculture through mechanization, artificial fertilizers and enhanced plant breeding (see Chapter 4).[3]

In many ways, this was a wonderful thing, and in better-off countries in particular, many people have enjoyed longer lives with less disease and higher standards of living. Growing population and increasing economic activity, however, required land and other natural resources. For this reason, the explosion of human numbers and wealth since 1800 has had devastating impacts on the living world, with natural environments and species rapidly being replaced by agriculture and other human activities with accompanying damage to ecosystems.

Setting for the Start of Farming

From the nineteenth century on, geologists have employed a variety of techniques—many derived from biology, chemistry and physics—to learn a great deal about the conditions that prevailed on the earth in the distant past. Probably the most famous of these involve paleontology, the study of fossils, although even fossil-free rocks, glacial ice and atmospheric gases have yielded significant information.

One surprising pattern they discovered that earth's history includes five extensive periods of glaciation when the planet cooled down enough to develop large areas of persistent ice fields as the result of the slow compaction of extensive, perennial snow fields.[4] These were especially common at higher latitudes and higher elevations. The first four lasted from tens to hundreds of millions of years. Geologists believe that all were characterized by alternating warmer and colder times with expansion and contraction of existing glaciers, especially in mountains and near the poles. About 2.5 million years ago, however, around the same time that our Australopithecine ancestors were coming to rely more and more on the stone tools they had learned to make, a major new episode of glaciation developed (see Chapter 1), and since then glaciers have swelled to cover large areas of the continents and then shrunk back at least four times. Throughout nearly all this time, human ancestors lived as small populations of hunter–gatherers, contracting and expanding their ranges as glaciers alternately covered and exposed the land.[5] The last great glacial retreat which is also a marker for the end of the Pleistocene period of earth history occurred about 12,000 years ago, and during the Pleistocene's hundreds of thousands of years, ancestral human populations were small—almost certainly less than ten million.[6]

Throughout the majority of this most recent glacial epoch, the human branch of the tree of life was represented by a number of different species, some replacing one another, and others coexisting through time. Neanderthals, for example, as well as perhaps two other species of humans survived alongside our own ancestors until thirty to fifty thousand years ago.[7] During this period, hunting tools and cultural knowledge about stalking and killing prey continued to advance, with big game animals increasingly succumbing to hunters (see Chapter 1). By around 11,000 years ago, as we have already seen, the majority of the largest land animals had disappeared, while Neanderthals and the other late-surviving human species were gone as well. Not long after, global climate experienced irregular oscillations between warmer and cooler periods, and it seems likely that the combination of these two circumstances—the hunting out of the largest land animals, and the onset of warmer, drier periods—were important spurs to the development of early farming.[8] Farming and herding began long before written records, so their beginnings can only be worked out from the archaeological record.

In the twentieth century, archaeologists discovered and explored many ancient sites, and from these it is clear that agriculture of one kind or another developed independently in at least four locations between twelve and five thousand years ago—in the Middle East, China, South America and New Guinea. The very first farmers were probably hunter–gatherers who, as foragers, had collected the seeds of wild grasses in the Near East. Between ten and twelve thousand years ago, some of them intentionally began saving some of the wild seed they had harvested in order to plant it during the next growing season on ground that had been intentionally burned over. Their goal was to provide a larger and more predictable harvest than wild plants alone could provide. Their first crops seem to have been grains like wheat and barley, and these were later followed by legumes like peas, lentils, and beans, foods relatively high in nutritional protein, as well as flax for fiber. In tropical New Guinea, early crops included yams, bananas and sugar cane.[9] It also seems clear that early farmers still continued to do some foraging and hunting for wild plants and animals even after they had begun to practice cultivation.

Transition to a Farming Way of Life

Hunter–gatherers were expert naturalists and constant observers of the natural world; they were well aware that many plants grew best when they received ample sunshine and were not closely surrounded by other vegetation. Out of this came the understanding that to grow abundant crops, natural vegetation generally had to be cut out or burned off. Such *clearing* of course, tended to destroy the existing natural communities that had occupied these locations for hundreds, thousands or even millions of years. In many cases, fire alone was the chosen tool for eliminating the native plants; in other cases axes were used; sometimes both techniques were employed.

The dawn of agriculture was a series of small-scale events, and though evidence of the particular food plants that were grown has been found again and again in archaeological sites, there is no good way to be sure of the extent of land cleared. In any case, the rapid loss of nutrients from cleared, cultivated soils may also have led to shifting cultivation, a practice sometimes known as "slash and burn." In such a system, relatively small areas of native vegetation are cut and burned, producing a residue of nutrient-rich plant ash. Crops are then grown on the site for just a few years before the cultivators abandon the spot and let it grow back up into native vegetation. At that point, they move on to a new patch of naturally vegetated land, cut it, burn it, and again grow crops for just a few years.[10] Two factors made this system reasonably sustainable in the early days of farming: first, cultivation of a given patch did not go on long enough to massively damage the soil; and second, nearby surviving patches of native vegetation retained wild plants, their seeds and the necessary animals to disperse them into the former clearing. Of course, this was only possible as long as farmers' numbers were small, cultivated areas were tiny, and areas of naturally vegetated land still dominated the landscape. Over time, however, shifting cultivation became impractical in most places due to widespread land clearance, permanent settlements and the pressure of increasing human populations.

The earliest gardens and small agricultural fields were probably located in the middle of the natural habitat of the wild ancestors of the cultivated plants—where the wild wheat grew well was almost certainly the kind of location where people cleared and then planted the first small fields of the grain. As farming became more widespread, however, people also learned that many different food plants could flourish in a few locations where soils were unusually deep and rich, often along rivers and in other lowland areas. In many places—and this is still very much the case today—these habitats became the focus of farming efforts, with the result that the plants and animals that were typical of such fertile, lowland places experienced disproportionate losses of habitats and species, with the most specialized and localized plants and animals being the first to disappear.

Around the time that people started to cultivate crops, they also started domesti-cating the first grazing animals—sheep and goats—which provided many useful products such as milk, meat, wool, hide, horns and manure for fertilizer.[11] Where domestic flocks were small and spread out, this may not have represented too much of a change from the natural grazing of the livestock's wild ancestors. As human populations grew, however, and the number and size of their flocks expanded with them, over-grazing and trampling became increasing sources of damage to natural grasslands and hillsides. Eventually, as agriculture and herding continued to expand, large areas of the natural landscape of the Near East and other early agricultural zones experienced profound change, with some native plant and animal species disappearing completely and others experiencing reductions in their abundance and distribution.[12]

Farming and herding allowed human numbers to increase. For thousands of years, however, people also continued to do some foraging, especially in the upland areas where soils were often too thin to support good crops, and in wetlands, where open water and waterlogged soils also prevented cultivation. Hunting and gathering had always had some impact on native populations of plants and animals. But except for the slowly reproducing big game animals which our ancestors seem to have over-hunted to extinction in many parts of the world, many animal and plant populations were still able to maintain themselves in the face of low levels of hunting and gathering. As agriculture slowly expanded human food supplies, however, and our numbers gradually grew, the number of potential part-time hunter–gatherers foraging on the outskirts of agricultural areas significantly increased, often resulting in further declines in wild populations adjacent to agricultural settlements. This pattern was also made worse by the expanded harvesting of trees for firewood and building materials. Even in areas where bricks formed the basic construction material, wood was still commonly used not only for fuel, but also for building posts, beams and roof timbers.

Environmental Impacts of Land Conversion

The growth of agriculture from its onset to the present day has usually involved the "clearing" of land. For this reason, it is critical to examine in detail some of the environmental changes that result from this process. A good way to begin is to consider some of the many things that are going on within living communities that have not been cleared. Depending on the warmth, wetness and seasonality of the climate, and on the underlying rocks which contribute to the formation of soil, very different kinds of plants generally come to dominate in different regions—water-hoarding succulents like cactus in the desert, tall hardwoods covered with vines and air plants in tropical wet forests, conifers like spruce and larch in cold, northern forests, and grasses and herbs on prairies and steppes. Each climatic region supports a fairly stable type of vegetation architecture, although the actual plant species living in the community will vary from place to place and continent to continent. Natural disturbances do occur—storms, lightning fires, floods and droughts are common causes—but significant damage is usually localized, and the surrounding areas that are left reasonably intact can aid recovery by providing seeds, spores and replacement animals.

Soil and the plants that grow in it are continually interacting in many ways, both directly and through the effects of the many kinds of bacteria, fungi and other small organisms that are always present. Most notably, plants are continually dropping leaves and other parts which fall to the ground to rot in place where they release their nutrients, although in some drier climates, lightning-set fires are necessary to liberate the mineral nutrients present in the plant litter. In any case, most of the nutrients present in dead plant material end up locally in the soil, while the vast networks of plant roots and fungal threads join with algae, bacteria, lichens and sticky humus to effectively hold the soil together even in the face of heavy rains and strong winds. The result is effective and continuous recycling of plant nutrients which are capable of moving back and forth between the soil and the plants almost indefinitely.[13] Soil is also a vast storehouse for the seeds and spores produced by the local plants, a key factor in allowing vegetation to regenerate following disturbance.

Ecologists often think of an area's vegetation, soils and microbes as forming the foundation of natural communities. But over hundreds of millions of years, animals too have come to be essential components of land habitats, developing intimate and intricate relationships with both plants and microorganisms. It is not, in fact, unusual for thousands of different native species to be present in an intact habitat, a number which can rise into the tens of thousands in the richest tropical rainforest environments. For this reason, a full scientific description of a given natural environment generally needs to include the climate and soils plus all the plants, animals and microorganisms, as well as all of their interactions.[14]

From the nineteenth century up to the present, biologists have carried out thousands of surveys of the living things present in different natural habitats. They have discovered a great deal based on visual inspection, basic measurements and simple chemical tests. In recent decades, however, they have also been able to document many of the barely visible changes going on in wild populations as they attempt to survive through time. Through their constant interactions, individual species continually impose genetic selection pressures on one another: plants not only photosynthesize but also make hundreds of substances which deter feeding by herbivores; animals attack both plants and other animals for food; decomposers break down dead organisms; parasites eat and live on other organisms. When these interactions are multiplied across all the species living together, it becomes clear that there is a gigantic set of established relationships in most natural habitats in which creatures interact not only behaviorally, but also chemically and physically through feeding interactions and through plant competition for nutrients, light and water. And this network of relationships also influences the way natural selection molds different species' genetic makeup. The overall complexity of these living mechanisms of *co-adaptation* of all the species in a community is hard to overstate— the sum total of the genetics of an entire environment of living things, especially in rich tropical communities, is mind-bogglingly complex.

In addition to the purely local effects, some interactions also occur regionally. Mobile animals like birds and some mammals may move nutrients, seeds and spores from one place to another. Water flowing downhill in watersheds and wind-blown dust can also contribute to these regional exchanges. All of the biological elements involved—everything except the non-living matrix of the rocks and soil—are also underlain by their genetic makeup which exposes them to natural selection and allows them to change over time.

When large areas of natural habitat are permanently destroyed—whether by cutting, burning, bulldozing or paving—the most visible impact is the mass destruction of the animals and plants native to that place. Where before there had been a forest or a grassland with many species of native plants; now, nearly all the native plants are gone. Where earlier there had been dozens or hundreds of different kinds of animals living on and among the plants, many have now been wiped out. Disappearances like these can readily be seen, but important losses also occur in the thousands of invisible creatures that used to be at work in the soil— the microscopic animals, bacteria and fungi.

Following clearing, the soil environment also rapidly begins to change. Organic matter that continues to decompose is not necessarily replaced; its loss allows the mineral particles in soil to more readily wash or blow away under the impact of rain and wind. Important plant nutrients are also likely to disappear as phosphorus and potassium dissolve in precipitation and runoff, while nitrates either wash away or are degraded to nitrogen gas, which is lost to the air. The sudden addition of these water-soluble nutrients to streams, rivers and lakes often causes serious pollution.

Once agriculture had begun, destruction of pre-existing plant communities was often undertaken with a specific use in mind—agricultural fields, cattle raising, forestry plantations and so on, activities which generally required the introduction of a small number of non-native species of economic value accompanied by other exotic species that arrive by accident. Eventually, these introduced species and their camp followers will establish new soil profiles and patterns of nutrient flow. In almost all cases, however, the new cattle ranches, oil palm plantations or corn fields are ultra-simplified ecosystems which contain far fewer species than were present in the natural environments they displaced—in some cases dozens or even hundreds of times fewer. And unlike intact plant and animal communities which are often unique because they have developed over millions of years in isolation as the species they contain adapted to one another and to the specific climate, geology and geographical isolation of their particular part of the earth, the new land uses which follow clearing are almost always the same as those in thousands of other places.

In fact, in order to serve their economic purpose, this land is managed to make it as similar as possible to other cattle ranches, oil palm plantations or corn fields. In this way, the astonishing richness of nature, built up over hundreds of millions of years in response to subtle variations in earth's biology, geology, geography and climate, is banished from cleared land, to be replaced with endless repetitions of highly simplified living systems which often can only be maintained by energy-hungry machinery in combination with repeated inputs of synthetic fertilizers, insecticides and herbicides. Cleared land then generally loses not only most of the visible plants and animals it used to support, but also many of its invisible riches—its microbes, nutrients and stockpile of seeds and spores. In many cases, variety is replaced with repetitive uniformity.

Landscape Ecology and Habitat Fragmentation

In most parts of the world, the surface of the earth is not uniform. It has high spots that are usually drier, and low ones that are usually wetter. There are hills, cliffs and rocky outcroppings many of which have a shadier and a sunnier side. There are thicker and thinner soils, and especially in areas that have been covered by glaciers, soils of different texture, depth and chemistry. For this reason, in many places when marshes, grasslands or forests are converted to agriculture, smaller, less desirable patches are often left out of the clearing process because they are not economical to cultivate. Sometimes the result is a patchwork of areas left in a more or less natural state. Some people view these as accidental nature preserves, although there are, unfortunately, many strong limitations on their conservation value. Compared to the much larger living communities out of which they have been carved, surviving habitat fragments have lost a great deal of their natural value. Being relatively small, many are usually missing some of the original regional species in the first place, while the population sizes of the species they retain are abnormally low. Surrounded by large areas of cleared land, their micro-climates are altered, generally making them drier and windier as well as increasing their sun exposure around the edges. We have already mentioned the large number of *tag-along* exotic species that generally accompany the plants and animals selected for use in agriculture. Some of these usually spread into surviving natural habitat patches, and a handful of them often become aggressive invaders, actively crowding out native plants and animals that remain in the fragments.

The dramatic declines in numbers of wild animals and plants following clearing have a range of serious impacts. When numbers fall sharply, populations generally lose some of their original genetic diversity. This means that they have a narrower range of solutions to problems of survival, while the likelihood of undesirable inbreeding increases. In this context, it is important to remember that most living things carry within their cells two copies of every gene, one each on matching stretches of paired chromosomes. Because of this, the harmful effects of a single bad gene can often be overcome by relying on the matching good gene on the other chromosome. When populations crash, however, inbred mating often occurs between closely related individuals who are more likely to be carrying around very similar genetic defects. The result—documented again and again in wild creatures, and even in some human royal families where marriages have taken place between close relatives—is a dramatic increase in genetic abnormalities and disease.[15] And when this happens, already reduced population numbers are likely to fall even further due to sterility and birth defects, further reducing a population's overall ability to respond to future environmental changes.[16]

In the nineteenth and twentieth centuries, for example, Florida's populations of the North American panther, or mountain lion, are estimated to have been reduced by 95 percent due to hunting, collisions with automobiles on networks of new roads, and the fragmentation of natural plant communities through agricultural development. By the 1970s it was estimated that only twenty panthers remained in the wild in south Florida, a situation that led to severe inbreeding. Conservation efforts since then have brought the number of animals up above one hundred, but the effects of inbreeding remain, and the surviving individuals have an elevated frequency of heart problems, weakened immune systems and, among the males, undescended testicles. There is also some evidence that male panthers have been feminized by exposure to agricultural chemicals.[17]

These are the short-term genetic impacts of habitat fragmentation, but the breaking up of habitats into discontinuous patches also worsens populations' genetic makeup in another way. Where habitats are relatively continuous, there is usually some degree of natural out-breeding caused by the movement of creatures and their genes through space—biologists label this *gene flow*. The seeds and spores of plants and fungi, for example, are often carried by wind, water and animals over a surprising distance. And in the case of animals, exploration, food-seeking and wandering away from high concentrations of their own kind often result in some individuals mating at some distance from where they were born. When this happens, it has the effect of spreading genes through space and of maintaining a healthy level of out-breeding throughout a species' range.

Following fragmentation, however, small, isolated populations may no longer be able to take advantage of this. Many woodland creatures will not cross open farm fields or even roads; this means that genetically speaking, they can end up isolated on very small habitat *islands* where the dangers of in-breeding and local extinction are very high. Local extinctions also sometimes occur in more continuous habitats, especially after storms, fires and the like. But when this happens, connections with other patches in which disaster did not strike makes it possible, and even likely, that missing species will eventually recolonize the damaged areas. This, in turn, gives the overall population of non-fragmented species both greater capacity to recover from downturns, and more long-term ability to use new combinations of genes to respond to changes in their environment. For a number of reasons, then, small, isolated populations are much more vulnerable to local extinctions than are larger, connected ones, and species left behind in small fragments are much less likely to be able to survive when faced with challenges like a drought, a harsh winter, increased competition or an outbreak of disease.

Before the many science-based advances which created modern agriculture, people tended to grow crops mostly in areas where soils, climate and topography were most favorable. Originally, at least, much of the earth's surface was too cold, too wet, too hot and dry, too steep—or in the case of many tropical rainforest regions—too conducive to plant, animal and human diseases to make human survival based on agriculture a viable option. Given these facts of geography, the development of farming and grazing was initially concentrated in a limited set of earth's environmental zones. These included temperate lands receiving modest amounts of rainfall whose natural vegetation was grassland, shrub land or savanna, along with selected temperate and tropical forest areas which received somewhat higher rainfall.

Given the early concentration of agriculture in such places, it was natural that they would experience disproportionately heavy losses of natural diversity. Historical records and food debris from archaeological sites give us a few glimpses of the biodiversity that disappeared as farming and herding spread, but agriculture went on for thousands of years before any written records were kept. In the longest settled, most heavily cultivated or grazed places, especially in the Old World, many species, perhaps even hundreds of thousands, almost certainly disappeared as fields and rangelands spread, with the greatest losses occurring in those places where the original natural communities had been the richest.[18] Along with the loss of living communities, the complex, balanced natural flows of water, energy and nutrients that had developed over long periods of time were also disrupted.

The advance of science and technology in the eighteenth and nineteenth centuries dramatically accelerated increases in farm productivity. For thousands of years before that, however, advances came very, very slowly with food production remaining stuck or even declining due to loss of soil fertility during long periods between bouts of innovation. Under these conditions, it took very little to push people's diets over the edge into inadequacy, and in China, for example, one of the first countries to support very large numbers of people through intensive agriculture, *hundreds of famines* were experienced over the past 2,000 years.[19] Nor was Asia the only continent afflicted by severe human food shortages. In Ireland between 1845 and 1852, failure of the staple potato crops due to plant disease is believed to have killed at least a million people, or more than ten percent of the population. Worldwide, the experience of hunger and deaths from food shortages in agricultural societies has also left a powerful psychological legacy which strongly devalues any practices which might interfere with potential food production. This undercurrent is part of the resistance to the setting aside of uncleared areas of land for purposes of conservation, to limitations on the drainage of valuable wetlands and to the uncontrolled use of pesticides.

Farm Technology

Many farming practices have changed beyond recognition since the first tiny plots of grain began to be intentionally planted ten to twelve thousand years ago. In many ways, however, the basic challenges facing farmers remain the same: finding the right plants to grow, usually from seeds or cuttings, and then planting, cultivating, irrigating and harvesting crops at the right time with the right techniques. Soil preparation and weeding have generally been the most laborious aspects of this work (see Chapter 6). In most natural plant communities, soils are generally loose and porous enough for water and air to penetrate and for roots to grow. But once soils have been farmed, they easily become compacted so that they require regular loosening in order to allow good conditions for root growth and the penetration of water and air. Competition from non-food plants (weeds, grasses, shrubs and trees) also has to be managed, and an adequate supply of plant nutrients has to be retained or periodically restored.

This is not a terribly hard routine to describe. But in the real world, it is a very challenging, physically demanding, and sometimes nearly impossible goal to attain, particularly given the uncertainties of weather, soil erosion, nutrient-depletion, plant-eating insects, weeds and the like. Even if a farmer does all the right things, at the right times, in the right places, success is never guaranteed. And even once the crops have been grown, there is still much to be done during and after the harvest—processing grain and other crops, storing them, and selecting and saving the best seeds for future planting. We will touch on some of these techniques in passing, particularly where important advances in farming technology have led to increased food supplies. But since our main concern is with the destruction of natural living communities, our initial focus here will be on land clearing followed by continuous cultivation, the single human activity which has caused the most harm to wild species.

Soil itself is made up of mineral particles like clay, silt, sand and stones, mixed with air, water and compost from decomposed plants and animals. Of these ingredients, the mineral particles present the greatest challenge to cultivation. They are heavy, abrasive and often packed together and so tend to rapidly wear away wood, bone or soft metal tools which come into contact with them. For this reason, the history of farm tools is partly the story of the development of harder and harder materials which could better resist the wear and tear of digging (see Chapter 4).

The archaeological record of the implements used in early farming is poor. Animal bones and sharpened sticks were almost certainly the first tools used for cultivation, although both tended to wear down pretty quickly in the soil. Eventually, the long human tradition of axe making (see Chapter 1) led to the production of stone-headed digging hoes which were strong and heavy but tended to chip or shatter in rocky ground. By about 7,000 years ago, simple wooden scratch plows had begun to be used in the Near East, a development probably related to the domestication of cattle—oxen were strong enough to drag a plow through the earth—although even very hard wood was rapidly worn away by continued contact with all but the softest soils. Among metals, lead and copper were both known by that time, although both were far too soft to make good farm implements. Only with the development of bronze about 6,000 years ago—a strong alloy of copper and tin—was a material found that could dramatically increase the durability of farm implements.

Bronze's two constituents, however, were both uncommon or rare in the earth's crust.[20] Copper's earliest uses had drawn on deposits of the naturally occurring pure metal which was only found in a handful of places around the earth. By about 7,000 years ago, however, people figured out how to extract it from ore rocks by heating them at temperatures of roughly 2,000°C—one theory is that this first occurred by accident in a pottery kiln.

Bronze was much harder than copper, and where its expense was not prohibitive, people also soon began to use it for knives as well as a wide range of other tools and weapons, some of which were used in agriculture. Compared to plows made entirely of wood, bronze blades or plow tips allowed deeper cultivation, a capability that was particularly significant in long-used fields where repeated cropping had significantly reduced the supply of nutrients in the top-most layers of soil. Bronze's major drawback was its high cost, which put it beyond the reach of most poor farmers who instead continued to rely on rapidly wearing wooden plows.

The use of bronze for hard-wearing tools went on for several thousand years. Around 1200 BC, however, metal technologies advanced to the point where it became possible to make durable farm implements out of a different material— iron. This element is one of the most abundant in the earth's crust, although it is harder to refine than copper and requires temperatures of 2,800°C to produce pure metal from ore. Iron tools began to made and used in the Near East about 4,000 years ago, and by 2,500 years ago, people in China had even begun making iron plow blades, although many of the early ones were not significantly harder than bronze. Over time, however, blacksmiths discovered various ways of shaping and hardening iron and began to make increasing numbers of affordable, useful implements which were almost as durable as bronze yet cost dramatically less. For a long time, poor farmers still had to rely on plows made entirely out of wood, although in many places, covering plow tips with iron significantly increased their durability, a practice which was widely used in the Roman Empire. One key factor, which did not really become well understood until the nineteenth century, was that iron that contained a carefully controlled amount of carbon as an alloy was much harder than pure iron, and in fact constituted a new alloy which came to be called *steel*.

From the 1700s on, steel became more widely used, although it wasn't until the innovations of Henry Bessemer in the nineteenth century that large volumes of inexpensive steel became widely available, providing a material for plows and other farm implements that was both harder and less brittle than the previously used cast iron. (The first high-production steel plows began to be made in the late 1830s by a blacksmith named John Deere, the founder of a company that is still a major manufacturer of agricultural equipment today.) Steel plows permitted deeper cultivation even in previously uncultivated grassland soils that were packed with grass roots. As a result, natural grassland habitats, including the North American prairies, were rapidly converted to agricultural fields. This allowed dramatically increased food production which, in turn, supported growing human populations, but also led to severe soil erosion and catastrophic losses of native grassland plants and animals.[21]

Plows were probably the first *farm equipment*—as opposed to hand tools—and they allowed more efficient crop production. Combining them with draft animals encouraged the conversion of larger areas of natural forests and grasslands to agriculture—much more land could be cultivated with animal-drawn plows than by people with hand tools. The enhanced ability of the plow to more fully loosen the soil, however, also increased the dangers of erosion—soil loss due to wind and rain. In native plant communities, erosion had generally been very limited except along the banks of streams and rivers, because living and dead plant roots were always present to anchor the soil. (Plant roots are covered with microstructures designed for water and nutrient absorption; these have an enormous surface area in contact with soil particles and so tend to strongly hold the particles in place.) But on the North American prairie, for example, erosion caused by plowing, wind and water has removed as much as half of the original topsoil in just a century and a half. Estimates for Iowa showed that pre-cultivation topsoil was 12–16 inches thick, but that by the late twentieth century it had been reduced to only 6–8 inches.[22]

By the time of the Roman Empire, simple, animal-powered reaping and threshing machines had also been invented, along with metal versions of most of today's common agricultural hand tools—forks, spades, sickles, scythes, shears, axes and saws— although back-breaking labor by people and animals still remained the norm.[23] In the Roman world the draft animal of choice was the ox. Horses were known, but the Romans had no effective way to harness them to pull heavy loads. (The first effective horse pulling hitch was invented in China around the second century BC and then continually improved until it attained its modern form by about the fifth century AD. Knowledge of the Chinese horse collar, however, did not reach Europe until roughly the tenth century, at which point horses slowly began to supplant oxen as *farm tractors*.) Before pulling collars were perfected, horses had still commonly been used by armies, but even in the military sphere, they faced the problem of damage and excessive wear to their hooves. The Romans coped with this through the development of strong leather sandals that were placed on horses' feet, although these proved to be far from ideal. It was therefore a major innovation when nail-on iron horse shoes came into use in Europe around the fifth century AD offering superior protection for horses' hooves and paving the way for horses' later selection as the draft animals of choice.[24]

Prior to the development of modern power and chemical technologies (see Chapters 6 and 4) improvements in agricultural technologies appeared slowly and in piecemeal fashion. Better plows, harnesses for draft animals, and rotation of crops were all advances. Around the year 800 in Europe, the two-field system in which half of farmland was left unplanted each year in order to recover some fertility, began to give way to a three-field rotation in which only one third of the land was left fallow. This required additional plowing and more wear on implements, but advances in iron production and blacksmithing made this increased effort feasible.[25] The three-field system also led to the planting of a wider range of crops including more legumes, plants which increasingly came to be recognized for their ability to actually improve soil fertility, although the explanation for this—the capture of atmospheric nitrogen by symbiotic root bacteria—only came to be understood near the end of the nineteenth century. In Europe at least, the late Middle Ages and early modern period also saw more destruction of woodlands and wetlands for agriculture, the first of which was accomplished by cutting and burning, and the second by draining. In both types of areas, large zones of natural habitat were utterly transformed with the loss of many of their native animals and plants.

The end of the fifteenth century marked the beginning of trade between Europe and the Americas, and within a few decades worldwide agriculture began to be transformed through the transfer of new crops between the Old and New Worlds. In what has come to be known as the Columbian Exchange, key food plants like maize, potatoes, squash, peanuts and tomatoes were introduced to Europe, Asia and Africa, while Old World staples like wheat, barley, rye, onions and apples moved west. In that same century, pioneering farmers in western Europe developed a four-field system in which wheat, rye (or barley), clover and turnips were continually rotated around, with farm animals being allowed to graze the clover, and the turnip crop being used for winter animal feed. Each field could now be planted every year and the increased animal food also meant more manure, which added to fertility.[26] Around this time farm machinery also advanced with Europeans beginning to design and make seed drills reducing seed waste and increasing farm productivity, although this technology was not really perfected until the nineteenth century. (Seed drills had first been first developed in China around the second century BC, but did not spread from there to other parts of the world.)[27]

The eighteenth century saw the beginnings of veterinary medicine which brought improved understanding of animal disease. It also saw the start of systematic livestock breeding which generally involved crossing of different breeds of domestic animals followed by some degree of careful in-breeding. Records were kept and measurements made, and the result was the development of more productive and consistent farm animals, which, in turn, provided more meat, wool, milk, etc. The global trade in fertilizer also increased with new supplies opening up including wood potash from Canada, a product made by cutting down and burning forest trees.

The nineteenth century saw even more profound technological advances in agriculture. Breakthroughs in mechanical and steam technologies allowed the creation of a range of new machines for plowing and other tasks, including several effective grain harvesting or *reaping* machines that appeared in the first half of the century as well as steam-powered farm tractors.[28] From the 1860s on, the development of mass-produced barbed wire made it economically feasible to create large fenced areas for livestock, making it more difficult for native wildlife to move normally, and encouraging the transformation from open range to more intensive enclosed grazing. Expanding railroads allowed fast, low-cost shipment of grain, livestock and other agricultural goods creating large national food markets and spurring larger capital investment in food production. The development of the science of chemistry also allowed a range of new fertilizers to become available (see Chapter 4).

Agriculture and Population

At the dawn of farming, less than ten million people inhabited the earth. But the ability to grow our own food gave people the means to slowly increase our numbers, and by about 1800—even in the face of setbacks due to famine, disease and war—human population stood at nearly one billion, a one hundred-fold increase over 10,000 years earlier.[29] It goes without saying that large-scale land clearing was also necessary to feed so many people, although few if any records of this were kept.[30] Overall, this period of human history laid the foundation for the human dominance of the earth, while simultaneously initiating the widespread destruction of biodiversity and the disruption of long-standing cycles of nutrients and of materials like soil and water. By 1800, significant amounts of nature that had existed before agriculture had already been lost, although if people had been able to stabilize their population at that point, there is a good chance that we would have been able to reach some kind of sustainable equilibrium with the rest of the living world.

Agriculture's unique expansion of the human food supply that underlay the one hundred-fold increase in population over 10,000 years could never be repeated—the earth simply could not support one hundred billion people. And yet, in the decades following 1800, an even more rapid kind of population increase began to emerge, raising our overall numbers to two billion by 1930. Even this, however, was just the beginning of the explosion in our numbers over the twentieth century. Between 1930 and 2000, the number of people on earth *tripled*. Twice as many people were added over those decades as had been alive in 1930. Obviously, agriculture had to be extended and more land cleared to accommodate such astonishing growth, and new farm technologies also played a role. But the underlying driver of this remarkable *population explosion* lay in advances in public health and medicine that dramatically lowered the human death rate, especially that of infants, children and women of child-bearing age.[31]

Modern Developments

In the 1850s and 1860s the Austrian monk Gregor Mendel had discovered some of the most basic laws of genetics through statistical analysis of the traits of tens of thousands of pea plants which he grew in his monastery's experimental garden. His results, however, were published in little-known scientific journals and it took more than three decades before the scientific community fully grasped the significance of Mendel's discoveries.[32] By the early twentieth century, however, scientific plant breeding had come into its own and started to develop much more productive, hardy and disease-resistant plant varieties. In Italy between 1900 and 1930, for example, the plant scientist Nazareno Strampelli used systematic crossing and selection of different varieties of wheat to significantly increase the amount of grain that could be grown per acre, allowing Italy to become self-sufficient in producing its most basic food crop.[33]

During the twentieth century, the pace of agricultural innovation quickened even more. Gasoline-powered (and then diesel-powered) tractors were easier to use than steam tractors, and they could be linked to other mechanical inventions to run devices like hay and straw balers, hydraulically powered buckets, scoops and forks, spray equipment, and even combine harvesters that could gather, thresh and winnow grain all in one machine. Improved pumps became able to move and pressurize water to allow large-scale irrigation, and where grid electricity was not available to run them, gasoline or diesel engines could keep the water flowing. Reliable, affordable steel pipes (and later high-tech plastic ones) made it feasible to irrigate ever larger fields, and new devices like center pivot irrigators allowed for much more even application of water. Ever advancing chemical know-how also led to more and more synthetic agricultural chemicals (see Chapter 4). Other new applications of internal combustion technology included bulldozers, log-haulers and other land-clearing machinery, and these, along with the chainsaws that were invented in the 1920s, made it much easier to cut down forests and convert them to pasture or plow land.[34]

Initially, at least, the pace of agricultural innovation was fastest in industrialized countries, as had been the case with Strampelli's work in Italy. But with poverty, hunger and birth rates highest in less developed countries, around mid-century non-profit humanitarian organizations began to undertake initiatives designed to introduce sophisticated twentieth-century agriculture into developing nations like Mexico, India, Pakistan and Turkey. Central to this was the breeding of more productive, more disease-resistant varieties of key crops, particularly staples like wheat and rice. To produce bumper crops, however, the new varieties generally needed more fertilizer, and also in many cases more water. Eventually, a variety of agencies and groups, including the United Nations, set up programs designed to spread the new crop strains around the world, along with the fertilizer, irrigation techniques and know-how needed to grow them successfully.

The results were dramatic, with crop production in places like India and Pakistan roughly doubling within a few years, and eventually allowing those countries to become reliably self-sufficient in grain production for the first time even as their populations continued to rapidly grow. By the 1960s the results were being hailed as a *green revolution* and the plant geneticist who was at the center of it, Norman Borlaug, received the Nobel Peace Prize in 1970. In 1960, world population stood at three billion. But just fifty years later, around 2010, in the most explosive population increase ever experienced by human beings, it had more than doubled to nearly seven billion. As many people were added during those fifty years as had even been alive in 1960.[35]

From a humanitarian point of view, the green revolution was largely a success. From an environmental point of view, though, this intense application of technology to agriculture had numerous serious shortcomings, some of which could be reduced over time, and some of which could not. Most striking in the early days were the deaths and illnesses among farm workers resulting from the pesticides that were often used as part of the new techniques. Many researchers believe that even the less toxic pesticides which started to be used later have serious impacts on human health and on native animal and plant populations (see Chapters 4 and 5). In some areas, increased crop productivity also made expansion of farmed lands economically appealing and this, in turn, led to further destruction of forest or grassland communities, a process that is still continuing in several parts of the world especially to create oil palm plantations. Overall agricultural use of energy, mostly derived from fossil fuels, also tended to increase, especially for irrigation, as did water pollution resulting from runoff of fertilizers and pesticides. Local biodiversity tended to fall since farming focused on repetitive plantings of highly in-bred plant varieties. (The broader impacts of expanded chemical and energy use are explored in Chapters 4, 5 and 7.)

★　★　★

Biomedical Technology Spurs Population Increase

For most of human history, right through the twentieth century in some countries, infectious disease often caused by poor sanitation was the leading cause of death prior to old age, with women of child-bearing age and infants and children being particularly vulnerable. In India, for example, the overall death rate around 1900 was a sky-high 44.4 per thousand, whereas by the year 2000, it had fallen dramatically to just 6.4 per thousand.[36] Such dramatic falls in death rates were one of the major causes of the unprecedented growth in human population which occurred during the twentieth century.

Historically speaking, the diseases which caused the most deaths and did the most to slow the growth of human population were either viral or bacterial. Among those caused by viruses, smallpox was perhaps the worst, although influenza, measles, mumps, yellow fever and AIDS have also taken their toll—the influenza pandemic of 1918 alone, for example, is estimated to have sickened 500 million people and killed at least 50 million, a number equal to several percent of the world's entire population at that time.[37] Diseases caused by bacteria that killed enough people to change human history included plague, tuberculosis and cholera. By the second half of the twentieth century, however, human vulnerability to this catalogue of diseases had largely yielded to the power of medical science and public health technology, setting the stage for the greatest explosion of human population the world had ever seen.

Smallpox

Smallpox is an illness that spreads directly from person to person; it was first recorded in India, China and Egypt as long as 3,000 years ago.[38] By the eighth century AD it had created a horrific epidemic in Japan that is thought to have wiped out more than 30 percent of that country's population.[39] Over the subsequent thousand years, its scourge spread little by little around the world, following the patterns of human trade, migration and colonization. Its impact is hard to overstate, with different authorities suggesting that it typically killed between 20 and 60 percent of those infected, and that mortality rates in infants could be as high as 80 percent. It is also believed to have been the main killer of the indigenous peoples of North, Central and South America following contact with Europeans. Without a history of exposure and therefore no immunity to this Old World disease, many authorities believe that it caused the deaths of up to 90 percent of the New World's native peoples.[40] Even in Europe, where it had long occurred, it is estimated that in the 1700s it killed 400,000 people each year out of a population of roughly 175 million. Over the course of the twentieth century, out of a much larger human global population, it is believed to have caused between 300 and 500 million deaths.[41] In addition to its catastrophic rates of mortality, up to a third of survivors were left blind while many others bore disfiguring scars for the rest of their lives.

Those who survived smallpox's devastation could not get it again—they had lifelong immunity, and this eventually led to experiments at prevention through the deliberate exposure of children and adults to material taken from previous victims—a procedure that came to be known as *variolation* after the Latin name for the disease's sores. This was probably first practiced in ancient China or India, the idea being that it would produce a mild case of the disease which would then be followed by lifetime immunity. By the 1700s, this procedure was also introduced into Europe where it remained controversial because as many as two percent of those inoculated in this way developed cases of the disease severe enough to kill them.

By the 1760s interest had arisen in the possibility that exposure to a related disease called cowpox might confer immunity to the deadly smallpox. A number of people had observed that milkmaids, who frequently developed cowpox, never seemed to get the more serious disease. Over the next thirty years, several people experimented with cowpox exposure as an immunization against smallpox, although it wasn't until the highly skilled physician and scientist Edward Jenner began a series of experimental trials in 1796 that the promise of *vaccination* (as opposed to variolation) became apparent. Jenner understood scientific method and evaluated the value of cowpox inoculation against smallpox through a two-step process. First he exposed a number of patients to cowpox, and then, after they recovered from that disease, to smallpox. By doing this he was able to prove that cowpox did in fact confer protection against smallpox, although a limited number of those he had treated did go on to develop mild cases of the more serious disease.[42]

Soon, Jenner wrote up a short report of his findings which he called "vaccination" because of its connection to cows (Latin *vacca*), but it was rejected for publication by Britain's Royal Society as lacking sufficient evidence. In 1798 he therefore paid for the printing of a pamphlet describing his findings with twenty-three patients entitled *An Inquiry into the Causes and Effects of the Variolae Vaccinae*. Once this appeared, some other physicians began to follow Jenner's procedure, which very slowly began to replace variolation in Europe and North America. There was, however, broad resistance within the medical community, and it subsequently took more than four decades before Jenner's home country of Britain fully accepted his procedure and banned the riskier method which used smallpox material. In fact, it was not until the mid-twentieth century that through international cooperation smallpox vaccination became nearly universal, in rich countries as well as poor ones, and this, in turn, laid the groundwork for one of the greatest advances in human public health of all time, the total elimination of smallpox from the human population.

Eradication of Smallpox

In a few wealthier countries, widespread and sometimes mandatory smallpox inoculation with the cowpox virus was enormously successful, so much so that by 1900 the disease was scarcely present in countries like Britain and the United States. In other parts of the world, however, poverty, ignorance and the persistence of traditional beliefs in the face of scientific understanding allowed it to hang on much longer, causing hundreds of millions of deaths in the first half of the twentieth century. By the 1950s, however, effective international efforts were launched to immunize enough people with the vaccinia vaccine (based on a close relative of cowpox) to allow the disease to be conquered in one country after another. By 1959, the parent organization of the World Health Organization (WHO) adopted a resolution to try to eradicate the disease globally, an outcome that had become possible due to advances in the science of epidemiology and to increased coordination, training of health care workers and spending on vaccines. The 1960s continued to see numerous smallpox cases, but by the 1970s, successful campaigns allowed one country after another to be declared free of the virus. Near the end of that decade, intensive efforts were undertaken to detect remaining pockets of the disease. By 1980, scientists could find no more evidence of it anywhere and so were able to announce that smallpox, the most deadly disease in all of human history, had finally been beaten.[43] Along with thousands of other advances in medicine and public health, the conquest of smallpox became a significant factor that spurred the explosive rise of human population since the middle of the twentieth century.

Microbes and Disease

While Jenner's work had focused on smallpox, by the nineteenth century, scientists had begun to expand their inquiries in efforts to understand the causes of the full range of infectious diseases. Up to that time, common explanations for their origins ranged from the *spontaneous generation* of disease-causing entities, to becoming chilled, to the ancient *miasma* theory—breathing bad air. As early as the 1600s, critical experiments disproving spontaneous generation had been carried out, but it was only in the middle of the nineteenth century that significant advances in optics and microscope design made it possible for scientists to clearly show that germs or microbes were the cause of most infectious diseases.

One pioneer in this research was the Frenchman Louis Pasteur. In the mid-1850s he had been approached by a wine manufacturer who was trying to understand why some of his wine turned to vinegar after extended storage. Pasteur accepted the germ theory and also knew that yeast was responsible for the fermentation of beer. To find an answer to the wine spoilage, he undertook a series of experiments and established that a tiny microorganism, a bacterium, was responsible for the souring of the wine. It had already been demonstrated by Nicholas Appert and others that food could be sterilized with heat. But boiling wine removed its alcohol and also altered its flavor. Using careful experiments, Pasteur then showed that if recently fermented wine was gently heated to about 50°C (122°F) any bacteria it contained would be killed or deactivated without reducing the wine's flavor or altering its alcohol content. Thus was born *pasteurization* of food, a process that has prevented great economic losses and saved many lives, while also confirming the germ theory of disease. Originally used primarily to preserve wine and beer, in 1886 it was suggested that it could also be used to reduce spoilage in milk, and by 1912 techniques for doing this had become widely known.[44] Eventually, it became clear that heat and other means could be used to deactivate or kill most bacteria, allowing the spread of disease to be reduced in many settings.

Even though in the 1860s Pasteur had not yet become famous, his confirmation of the germ theory of disease had begun to influence the scientific community and even to save lives, and one individual who learned of his discoveries was the British doctor Joseph Lister. Lister completed his training in medicine in 1852 and by the 1860s had become a prominent surgeon and teacher of surgery in Scotland. Lister was an extremely conscientious individual who was deeply troubled by the high rate of wound infections and deaths following surgery. Once he learned of Pasteur's work, he resolved to develop new surgical techniques to reduce infections. (Rates of infection following surgery were so high in his era that some people were actually calling for a total ban on body cavity and brain surgery.)

By the mid-1860s Lister had established a whole a new set of procedures based on disinfection using carbolic acid (phenol). The operative area and even the air in the operating room was disinfected before, during and after every operation; surgeons washed their hands in disinfectant before and after each operation and wore clean gloves. The results of Lister's innovations were remarkable. Life-threatening post-surgical infections like gangrene, erysipelas and pyemia all fell to nearly zero following rigorous application of Lister's system. Research had started to establish that germs rather than bad air caused the majority of non-degenerative disease, and once Lister found a way to kill the germs, surgery began, for the very first time, to be reasonably safe. These developments not only marked significant advances in medicine and in the reduction of human pain and suffering, but they also contributed to human population increase, especially when applied to women in childbirth.[45]

The First Vaccines

The decades following Pasteur and Lister's pioneering work were marked by dramatic progress in the understanding of microbes and the diseases they caused with much of the earliest work being done in domestic animals. In Germany, a young physician named Robert Koch, for example, turned his attention to the study of the important livestock disease anthrax, and by 1875 he had managed to isolate the specific microbe which caused it.[46] The following year he was able to very accurately characterize the bacterium involved, *Bacillus anthracis*, and his proof that this microbe actually caused the disease represented the most complete confirmation of the germ theory up to that point.

In France, Pasteur too had long had an interest non-human diseases of economic significance—the spoiling of wine and beer, death among silkworms, and important diseases of domestic animals. Of the latter, he first studied fowl cholera, a disease of chickens, geese, ducks, turkeys and other birds. Well aware of Koch's recent work on anthrax, Pasteur benefitted in the late 1870s from a lucky accident as he was trying to isolate and identify the bacterium that caused fowl cholera. His lab had produced a culture of the germs taken from infected birds which was intended to be injected into healthy chickens to prove that it caused the disease. There was a mix-up, however, and the culture sat for a month before being tested on experimental birds. During those weeks, it somehow became less virulent, and though the birds injected with it developed some mild symptoms of cholera, they did not die. Later, he tried to infect the same birds with fresh chicken cholera cultures, but found that they did not become sick. Jenner had shown that smallpox could be immunized against by exposure to cowpox or by variolation, and now Pasteur realized that his less than fresh chicken cholera culture had produced a similar effect on chickens. These results not only further confirmed the germ theory of disease, but also suggested that it might be possible to confer immunity to other diseases through exposure to bacteria that had been modified to become less virulent.[47]

Anthrax and chicken cholera were economically important diseases, although they did not have major direct impacts on human public health—people could get anthrax, but it was not a common illness. By the 1880s, researchers like Koch had begun to focus their attention more and more on human diseases, and the first important illness whose cause was identified was the waterborne bacterial illness cholera. In 1883, after traveling to Egypt where cholera had recently caused numerous deaths, Koch managed to identify and isolate a vibrio bacterium which he believed caused the disease. He did not go on to develop a vaccine against cholera, but his findings laid the groundwork for another scientist, Waldemar Haffkine, who in 1893 did manage to create a cholera vaccine, the first effective immunization against a deadly human bacterial disease. After much trial and error Haffkine found that exposure of the cholera germs to blasts of hot air rendered them incapable of producing the full-blown disease, although still able to confer a reasonable level of immunity. Later that same year, Haffkine went to Calcutta, India where a serious cholera epidemic was underway and there he succeeded in immunizing large numbers against the disease.[48] Cholera killed tens of millions in the nineteenth and twentieth centuries, and as with other important immunizations, its decline due to vaccines, improved drinking water supplies and better treatments for those already sick became yet another driving force behind human population growth.[49]

Safe Drinking Water Supplies

Cholera was the first important human bacterial disease for which a successful vaccine was developed, but even several decades before that happened, scientific detective work had also led to critically important ways of preventing cholera that had nothing to do with vaccination. Serious epidemics of the illness had passed through Britain in the early 1830s and again in 1849, drawing the focus of the medical community. In 1850, several experts including a medical doctor named John Snow came together to form a new organization, the Epidemiological Society of London, specifically designed to scientifically investigate and counteract epidemics like cholera. Four years later London experienced another serious outbreak of cholera. Snow, who was highly skilled in public health as well as in medicine, used maps and statistics to show that one major cluster of cases had occurred around one particular public water pump—there was little indoor plumbing at the time. He then convinced the authorities to close that pump whereupon the neighborhood epidemic rapidly trailed off. He then went on to publish a detailed report showing that cholera was a waterborne disease that was almost certainly passed through contamination of drinking water and food by fecal matter from previous sufferers.[50]

Even though an effective cholera vaccine was not developed until the 1890s, in the aftermath of Snow and his colleagues' mid-century demonstration that drinking water had to be safe if infectious diseases were to be avoided, Britain and a number of other countries were able to dramatically reduce the incidence of cholera. The most important steps needed to ensure this were the installation of sanitary sewers designed to safely remove human waste from where people lived, the obtaining of drinking water from uncontaminated sources, and the creation of an adequate public health regime to carry out regular inspections and enforce compliance. In London, the original project to construct sanitary sewers lasted from 1859 to 1875, and that, combined with safer drinking water supplies, dramatically reduced the incidence of cholera and pretty much precluded the possibility of another major epidemic.[51]

Safe drinking water and sanitary sewers were not new to the world in the nineteenth century, although it seems that their value had to be discovered all over again. Ancient Rome had devoted considerable resources to building aqueducts to channel pure water from surrounding uplands to their cities, while also investing heavily in sanitary sewers. And even long before Rome, the prehistoric Indus Valley civilization in Asia had also made use of them.

Along with immunizations against infectious disease, the provision of safe drinking water saved countless lives, a trend which eventually was replicated in more and more countries around the world.[52]

Viruses

By the 1880s, the proof of the germ theory had led to greater and greater interest in the idea of immunization, and there was a general understanding that biological material associated with a given disease needed to be weakened to produce a vaccine. Louis Pasteur's laboratory was a center of this research and in the early 1880s, he and one of his collaborators, Pierre Roux, undertook to do just this for the terrifying disease rabies. Not usually spread from person to person, rabies was commonly contracted through the bites of dogs smitten with the disease. It attacked the nervous system, produced horrible symptoms, and was almost inevitably fatal. Soon, they prepared a vaccine derived from the dried spinal cords of rabbits that had perished from the disease, and Pasteur's group tested its effectiveness on fifty dogs. Then, in 1885, a nine-year-old boy named Joseph Meister was attacked by a rabid dog, and brought to Louis Pasteur. Though a chemist by training rather than a medical doctor, Pasteur decided to treat the boy with the anti-rabies vaccine, and the boy never developed the disease.[53] Practical progress was being made against rabies, but even with the best microscopes then available, even the most skilled microbiologists were never able to discover the microbe that caused the disease.

As the nineteenth century reached its final decades, however, the advance of microbiology continued to produce new techniques for separating and isolating bacteria from the solutions in which they were suspended. Among these were porcelain filters with holes so small that they would catch all the bacteria moving through them while letting the fluid surrounding them pass. This turned out to have at least two important uses: removing all bacteria from solutions; and isolating water soluble toxins that some bacteria produced and released into the liquid around them. They also turned out to be critical in the discovery of the other great class of disease-causing microbes: viruses.

Late in the 1880s, a number of researchers had turned their attention to the study of plant diseases, especially those affecting economically important crops. One of these was the Russian Dmitri Ivanovsky who studied problems with tobacco plants, most notably a *mosaic* disease which caused light and dark patches on tobacco leaves, slowing growth and often stunting young plants. Ivanovsky found that extracts from diseased leaves were still infectious even after passing through several fine bacterial filters, and in 1892 he published reports suggesting that some very small, unknown infectious agent was involved. His work was replicated in 1898 by the Dutch microbiologist and botanist Martinus Beijerinck who found that unlike bacteria, he could not grow this new disease causing entity in culture media, although he could get it to reproduce on living tobacco leaves. He therefore postulated that there existed tiny infectious agents, much smaller than bacteria, and he named them viruses.[54] From then on, viruses were of great interest to microbiologists, although the difficulty in studying them and their invisibility meant that progress in understanding them was very slow. (In 1911 Francis P. Rous showed that a different virus caused a particular cancer in chickens, thus demonstrating that viruses could cause animal as well as plant diseases.)

Electron Microscopes

In the late nineteenth and early twentieth century, physicists developed many kinds of experimental apparatus involving energy and particle beams, most of which ran on electricity. Many of their experiments involved the manipulation of beams of charged particles, and eventually it was realized that electromagnets could "focus" beams of charged particles just like glass lenses could focus beams of light. Electrons had dramatically shorter wavelengths than visible light, and some physicists began to think that it might be possible to make an *electron microscope* which would allow them to see things too small to be seen with the best light microscope. The idea was first put forth in the mid-1920s and the first workable instruments began to be produced by the end of the 1930s. By 1938, the enormous magnification possible with this new instrument led to the first definitive image of a virus—the ectromelia virus, which caused mouse pox.[55]

From then on, the ultra-microscopic world of infectious disease began to slowly open up to scientific researchers as instruments capable of magnifying hundreds of thousands of times began to reveal details that could be interpreted at the level of chemical molecules. (Light microscopes, in contrast, could at most magnify about 2,000 times.) Since then, thousands of viruses have been named and categorized, among the millions that are thought to exist. From the 1950s on, the emerging field of molecular biology, which grew out of discovery of the basic structure of DNA, has made possible ever-increasing understanding of viruses in particular and of infectious disease in general, allowing the creation of vaccines for serious viral diseases like polio, influenza, yellow fever, hepatitis and smallpox.

The advance of scientific understanding in recent centuries, and particularly since 1850, has revolutionarily transformed human health. Child and maternal mortality has fallen dramatically and the average human lifespan has been extended by several decades. For people, the reductions in loss, pain, fear and suffering has been incalculable, while the economic gains have been enormous. For the rest of the living world, however, it has been something of a disaster, as billions of extra people have claimed more and more of the earth's space and natural resources, reducing all other living things to narrower and narrower toeholds. It is even possible that science-based medical and public health advances, unless accompanied by much wider understanding and acceptance of science and ecological realities— including limitations on human populations—will push human numbers to completely unsustainable levels, leading to new sources of mortality from over-crowding, conflict, ecosystem collapse, resource depletion and disease.

★　★　★

Sailing and Navigational Technologies Lead to Trans-Oceanic Trade and Migration

So far in this chapter we have focused on the development of agriculture, the advance of agricultural technology and the progress made in understanding and combatting infectious disease. Along with the conversion of more and more land, these changes allowed human populations to sporadically grow until they reached about half a billion by 1500.[56] During the preceding centuries the vast majority of people had survived as poor farmers; there was little if any manufacturing, and any increases in the economic productivity of settled agriculture were continually being eroded by population growth. As European governments looked around, one of the few ways they saw to enrich themselves was through foreign trade, especially with east Asia whose large populations produced numerous unique and valuable natural products and finished goods. Trade across the Mediterranean, however, or by land across Eurasia, was difficult and expensive due to competing and often hostile political and religious interests. A number of people speculated that Asia could be reached by sailing west as well as by sailing east, but at the time there was still some uncertainty about the true dimensions of the earth and the location and extent of any continents beyond Europe, Asia and north Africa. In addition, sailing ships had trouble sailing close to the wind and navigational instruments were still in a primitive state.

By the early fifteenth century, however, this had begun to change, with Portugal in particular developing an ambitious program of exploration reaching down the west coast of Africa and out into the adjacent Atlantic. Part of the success of this program was due to improved technologies, especially navigational aids like the astrolabe and the cross-staff, and to a new and more maneuverable type of ship known as a caravel. Using these tools, the Portuguese reached Madeira in 1419 and the Azores in the 1430s, and soon Portuguese captains were also sailing farther and farther south along the African coast. By 1488 Bartolomeu Dias succeeded in reaching Africa's southernmost point, becoming the first to sail from Europe into the Indian Ocean. And once the possibility of such an extended, southern route to the Far East became established, it did not take very long for Portugal to dispatch a new expedition intended to make it all the way to India. And so it was that Vasco da Gama arrived on May 20, 1498 in Calicut on India's west coast, inaugurating a completely new trade route between Europe and Asia.[57]

In the decade between Dias' voyage to Africa's southern cape and da Gama's voyage as far as India, another Iberian expedition ventured out into the Atlantic in hopes of reaching Asia by voyaging west. Using a fundamentally flawed estimate of the distance to Japan if one sailed west, in early August 1492 Christopher Columbus led three ships westward from Spain. He took on fresh provisions in the Canary Islands and then reached the Bahamas in the Caribbean on October 12, still believing that his ships had arrived in Asia. Over the following decade, Columbus led three more expeditions to the Caribbean to explore and establish permanent colonies, bringing with him more than a thousand settlers from Spain along with their supplies.[58] This marked the start not only of countless trips back and forth between Europe and the Americas, but of numerous other voyages of discovery including the first circumnavigation of the earth between 1519 and 1522. By the 1640s, European sailors had also reached Australia and New Zealand, opening the way for continuing voyages among all the inhabited continents as well as to many remote oceanic islands.

These important events were part of the *Age of Discovery*, a period of several centuries in which progress in geography, navigation and sailing ship technology for the first time opened up the possibilities of regular global travel and global trade. It was also the beginning of far-flung colonial empires which at least in part were based on the large-scale transport of precious metals and other goods. Thousands of books have been written about the events of this era and its impacts on the lives of the hundreds of millions who either became subjected peoples or colonial rulers.[59] But our focus here is not on the cultural or economic effects of discovery and exploration, but rather on its complex and far-reaching effects on global ecology.

Trans-Oceanic Trade leads to Introduction of Exotic and Invasive Species

Understanding the impacts of the global movement of goods that accompanied human exploration and colonization requires a brief digression into historical biogeography. As European exploration between 1400 and 1800 reached all of earth's inhabited land areas as well as many uninhabited islands, the discovery of thousands of new species helped stimulate the development of modern botany and zoology. By the 1700s and 1800s, scientists were able for the first time to carry out systematic studies of the diversity of global life and the geographical ranges of thousands of different species. Between 1500 and 2000, hundreds of thousands, if not millions, of specimens of exotic plants and animals were collected and transported back to Europe and North America, where they were studied, described and preserved to form the core collections of those continent's natural history museums. Groundbreaking work on classifying them was carried out by scientists, the most famous of whom was Carl Linnaeus of Sweden, who published a series of extensive classifications of living things between 1735 and 1768.[60]

Arranging the thousands of newly discovered creatures from around the world based on similar characteristics was one thing, but figuring out how and where they had arisen was quite another. In Linnaeus' time no one really understood how different living things had developed in different parts of the world—Linnaeus himself still accepted the traditional explanation of divine creation. Over the ensuing decades, however, this little by little began to change as notable explorers during the first half of the nineteenth century, particularly Alexander von Humboldt, Charles Darwin and Alfred Russell Wallace, intensively explored South America and the East Indies. Humboldt spent five years (1799–1804) studying geology, botany, meteorology and geography in South America and the Caribbean;

Darwin nearly five years (1831–1836) traveling around the world as naturalist on HMS *Beagle*; and Wallace four years in the Amazon (1848–1852) followed by eight years (1854–1862) in the Malay Archipelago, where he studied animals and plants and made exhaustive collections of birds, insects and other life forms.

By 1859, all of this activity led to an extraordinary scientific breakthrough as Darwin and Wallace independently developed the first workable model of how new kinds of living things originated through a process of *natural selection*. By then, many researchers had noted that different continents contained very different kinds of animals and plants than could be found elsewhere—kangaroos in Australia were one of the most striking examples—and also that isolated islands often had their own unique species which could be found nowhere else. Advancing understanding of geology, soils and climate made it clear that the challenges facing living things differed from place to place, and natural selection now offered an explanation of why living species in South Asia, for example, were different from those in South America. In each place ancestral species had accumulated characteristics that fitted them to survive under the conditions present in their native locale in response not only to the climate but also to the challenges and opportunities presented by the other species that had also evolved there.[61]

Sometimes, as with the nectar-feeding New World hummingbirds and Old World sunbirds, groups of species evolved similar lifestyles and even similar appearances. Yet despite such superficial resemblance, detailed examination showed that they were not closely related. New World animals and plants, for example, had evolved on their own for tens of millions of years with little mixing with species from Africa, Eurasia or Australia, places thousands of miles away over the ocean. And similar patterns were seen all around the world. A rainforest in the Congo, and one in the Amazon might look superficially similar, and some of the same kinds of biological interactions between say insects and plant flowers would be going on in both. But the actual species of plants, animals and other living things would in most cases be quite different. These observations formed the basis of the science of biogeography, a field which was pioneered by explorers like Linnaeus, Humboldt and Wallace.

Later on, in the twentieth century, the developing science of ecology documented the tightness of the web of living interactions into which most species were bound as the result of this kind of shared local history. Animals, plants and even microbes had evolved in place over long periods of time with predators, herbivores, parasites and potentially damaging microbes. The result was that most species populations oscillated within fairly narrow limits, their potential to dramatically expand their populations controlled by effective competitors for food, light or water, as well as by creatures that might feed on them.

During the Age of Discovery, however, people began to intentionally move species from continent to continent on a large scale, sometimes to serve as food plants or livestock, sometimes—as with garden flowers—for esthetic reasons. Many other species also ended up being transported accidentally, as with the rats and mice which infested ships, and the weed seeds that were accidentally present in seed grain or in the soil of potted plants. Some transported exotic species failed to take;

others like the common lilac bush from the Balkans, grew well if planted but didn't spread widely without human intervention. A few, however, exploded in the new locations people moved them to, and increasingly they have become a major source of environmental disruption and damage. The worst offenders vary somewhat with the climate zone to which they are adapted, but a few worth mentioning are Japanese kudzu in North America, European rabbits in Australia, and African cane toads and giant land snails in many tropical areas.[62]

This problem of invasive species is not primarily a global tale, but rather a collection of thousands of local stories—what happened in Australia is different than what happened in North America, and what happened in Chile differs from what happened in British Columbia or even in Ecuador. Many different areas on the same continent often have different invasive species because of different climatic and environmental conditions, and different accidents of introduction. Some of the greatest damage occurred on oceanic islands—in the south Pacific, for example, it is estimated that as many as 1,000 species of birds, many of them flightless, likely went extinct as people colonized island after island bringing with them rats, pigs and other exotic species.[63] Predation or competition from these introduced creatures combined with human hunting and forest clearing to send the populations of many local native species plunging to zero.

Introduced animals often prey upon local species and also compete with them for food, nest sites and other resources. In the Everglades wetlands of south Florida for example, enormous Asian pythons introduced through releases by irresponsible pet owners have wiped out nearly all native mammal species in many areas, with some local populations reduced by 98 percent or more.[64] Invasive plants, free of the competing plants and the insects that typically fed on them in their original habitats, sometimes experienced explosive population growth which allowed them to either outcompete or chemically suppress the native species that originally were present. Locale after locale around the world has had its original natural communities damaged by one or more invasive plant and animal species, a serious unintended consequence of human migration, exploration, travel and trade.

In many ways, it is hard to educate the public about the problems caused by exotic and invasive species. Ever since the beginnings of agriculture, human civilization has been based on bringing along non-native species of plants and animals wherever people move or settle. Besides these hundreds of intentional introductions, thousands of other species have also accompanied our movements as stowaways in ships and planes or hidden in traded goods. Exotic species, especially those that become invasive, create a kind of biological pollution. Some do not do much harm; others, like the pythons in the Everglades and the species mentioned above, have the potential to critically damage entire ecosystems, some beyond the possibility of recovery. And in many cases, once they have been introduced, it is practically impossible to get rid of them.

★ ★ ★

Agricultural Development Over Time

We have already seen that the development of agricultural technologies completely transformed human life over the last 10,000 years. During the first half of this period, there were no written records, so human agricultural activities can only be documented through archaeological reconstruction. Later on, written records became more common—indeed, writing may have first been developed to keep track of agricultural production. In the second half of this period, however, documentary evidence provides some information about the amount and location of farms and pastures in certain regions, although many areas kept no records, and many documents have not survived. During the last three centuries, much more complete records have been kept about the extent of farm and pasture lands and the human populations they supported, and this has allowed scholars to more accurately describe the spread of agriculture and relate it to the ever-rising line of human population.

Prior to 1700, we can make only rough estimates of the exact extent of land cleared. What is known is that in a number of long-settled, heavily cultivated locations in the Near East, Egypt, China, India, southern Europe and Mexico, large areas of natural communities were either wiped out or reduced to remnant fragments. And this, in turn, led to species losses and widespread disruption of ecosystems, with drastic changes in pre-agricultural cycles of water, nutrients and energy. In many areas around the Mediterranean and elsewhere, enormous amounts of topsoil were washed away as a result of deforestation, over-grazing and farming, impoverishing the land. In other places, including Mesopotamia, early employment of heavy irrigation led to build up of salt in the soil to the point where the land eventually lost its fertility.

From the eighteenth century on, increasingly accurate agricultural records began to be kept, and more documents have survived, making it possible to develop more and more exact estimates of the total land converted from forests and grasslands to croplands, hayfields and pastures over time.[65] By the late twentieth century, the development of satellite technology also opened up an entire new chapter in the environmental analysis of human land use. Cameras and sensors operating miles overhead could now accurately measure the extent of land given over to forests, pastures, crops, urban areas, and a wide range of other land uses, while computers could compare the satellite observations with land use data gathered on the ground. Using this combinations of techniques, academic experts were increasingly able to develop highly advanced global land use estimates. What they found was that by the year 2000, the natural lands that people had converted for agriculture—including pastures and grazing lands— spread over 43,000,000 square kilometers (over 16,500,000 square miles) and covered an area more than twice the size of North America. This included 15,000,000 square kilometers (roughly 5,800,000 square miles) of farmland, and 28,000,000 square kilometers (around 10,800,000 square miles) of pasture and grazing land.[66]

How can we make sense of such a vast areas of natural communities that have been destroyed to make way for agricultural lands? One way, perhaps, is to see how many modern countries it would take to equal the total amount of land that has been cleared worldwide for farming and pasture. Skipping over several dozen of the smallest countries, and excluding those whose territory consists mostly of ocean, the amount of cleared crop land alone is equal to the *entire area* of the following 105 countries:

Albania
Armenia
Austria
Azerbaijan
Bangladesh
Belarus
Belgium
Belize
Benin
Bhutan
Bosnia & Herzegovina
Bulgaria
Burkina Faso
Burundi
Cambodia
Costa Rica
Cote d'Ivoire
Croatia
Cuba
Czech Republic
Denmark
Djibouti
Dominican Republic
East Timor
Ecuador
El Salvador
Equatorial Guinea
Eritrea
Estonia
Finland
Gabon
Gambia
Georgia
Germany
Ghana
Greece

Guatemala
Guinea
Guinea-Bissau
Guyana
Haiti
Honduras
Hungary
Iceland
Iraq
Ireland
Israel
Italy
Jamaica
Japan
Jordan
Kuwait
Kyrgyzstan
Laos
Latvia
Lebanon
Lesotho
Liberia
Lithuania
Malawi
Malaysia
Moldova
Montenegro
Morocco
Nepal
Netherlands
New Zealand
Nicaragua
North Korea
Norway
Oman
Panama
Paraguay
Philippines
Poland
Portugal
Qatar
Republic of China (Taiwan)
Republic of Macedonia
Republic of the Congo

Romania
Rwanda
Senegal
Serbia
Sierra Leone
Slovakia
Slovenia
South Korea
Sri Lanka
Suriname
Swaziland
Sweden
Switzerland
Syria
Tajikistan
Togo
Tunisia
Uganda
United Arab Emirates
United Kingdom
Uruguay
Uzbekistan
Viet Nam
Western Sahara
Zimbabwe

Total area of above countries = 15,189,096 square kilometers (5,864,505 square miles).[67]

Yet as we have seen, even 15,000,000 square kilometers does not represent the full extent of the natural communities destroyed or heavily modified by human activities. Beyond active farm fields and orchards, an additional 28,000,000 square kilometers of land has been converted to pasture or rangeland grazed by domestic animals, a change which severely reduces or even eliminates the plants and animals that were originally present, while also disrupting their natural ecosystems. These grazing areas on their own are equal in size to the 105 nations listed above, *plus* the following 21 countries:

Afghanistan
Botswana
Burma
Cameroon
Central African Republic
Chile

France
Kenya
Madagascar
Mozambique
Pakistan
Papua New Guinea
Somalia
South Sudan
Spain
Thailand
Turkey
Turkmenistan
Ukraine
Yemen
Zambia

Total area of these nations = 13,115,679 square kilometers (5,063,992 square miles). Total area of combined lists of countries = 28,304,775 square kilometers (10,928,235 square miles)

Beyond farms and pastures, roughly two million additional square kilometers of plant communities—an area larger than Venezuela and Namibia combined—have been converted to forestry plantations, sites where the original vegetation has been cleared and replaced, often with just a single species of fast growing tree, and frequently a species that is not even native to the country where it is now planted.[68] Tree farms do provide some environmental benefits through storage of carbon dioxide and limitation of soil erosion. When native tree species are used, there may also be significant benefit to wildlife. But overall, the biological benefits of forestry plantations are quite limited, since they usually do not support the overwhelming majority of native species that were present on the site before the original vegetation was removed.

The broadest perspective of all on the impacts of human land use can be seen by considering the total overall area of natural plant communities that have been destroyed or heavily altered through conversion to farm fields, pastures, rangelands and forestry plantations. Taken together, these amount to roughly 45,000,000 square kilometers, a vast area that would cover the countries in the above two lists, plus 16 additional nations:

Angola
Bolivia
Chad
Colombia
Egypt
Ethiopia
Mali

Mauritania
Mongolia
Namibia
Niger
Nigeria
Peru
South Africa
Tanzania
Venezuela

Total area of these nations = 15,319,892 square kilometers (5,914,699 square miles). Total area of combined lists of countries = 45,188,777 square kilometers (17,447,184 square miles).

Taken as a whole, converted lands make up an area equal to the total territory of 142 countries, and cover more than a third of earth's entire ice-free, land surface. Of course, not every inch of land in all the countries named above is actually under cultivation—many of earth's fields and pastures lie in the remaining nations not listed, including some very large ones like Russia, Canada, the United States, China, Brazil, Australia, India and Argentina. And some extensive areas of natural forest land still survive both in the far north, in nations like Russia and Canada, and in a few tropical areas like Amazonia. Those forests aside, though, most of earth's remaining non-agricultural land can be found only in extreme environments—mountains, deserts, and Arctic or subarctic areas.[69]

★ ★ ★

Life in Tropical Forests

During the second half of the twentieth century, several trends came together to create widespread awareness of environmental problems, and this led to national and international campaigns designed to preserve earth's dwindling biological heritage for future generations. In developed nations, rising standards of living led to greater emphasis on the quality of life beyond basic needs, while at the same time increasing economic activity and population growth around the world produced more and more pollution and accelerated the destruction of surviving wild communities. All these changes, however, did not go unnoticed. Environmental science had begun to make rapid progress, and ecology and other environmental fields were discovering and documenting more and more of the complex workings and biological richness of earth's surviving natural communities. But even as more of the remarkable details of natural systems and species were slowly emerging, the same advancing scientific fields were also repeatedly discovering evidence of the devastating impacts on natural systems and human health of air, water and soil pollution, toxic chemicals, forest clearance, soil erosion, over-grazing and over-fishing. Against this background, popular attention soon came to focus on the clearing of large areas of biodiverse, ancient forest in tropical countries where human population was growing rapidly.[70]

For millions of years before human agriculture, some of nature's most diverse communities were located in the wet, warm, frost-free regions known as tropical rainforests. These environments developed to their fullest extent in places like South America's Amazon and Atlantic Forests, central Africa's Congo region and on island and mainland areas of South East Asia, locations where climates were warm to hot year-round, with average temperatures hovering near 70 degrees, and annual rainfall generally greater than 100 inches. Apart from any high mountains located in them, these regions were completely free of frost, and their geographical location near the Equator meant that they had been relatively warm and wet for millions of years. These favorable climatic conditions were almost certainly one important factor in allowing the living richness of such communities to continue to accumulate over very long periods of time. As a result, these forests became perhaps the most complex and diverse habitats ever to exist on earth, a veritable treasure house of living creativity and adaptation, with plants, animals and microbes achieving their greatest variety of species and their most intricate adaptations to one another and to the physical environment.[71]

Though the long-term mildness of the climate is clearly a factor, ecologists have so far struggled to tease out precisely which conditions have been most important in the evolution of the wet tropics' astonishing levels of biodiversity. One hypothesis is that continuously favorable conditions allowed the overall process of evolution, including the appearance of new species, to proceed more rapidly than in temperate and cold regions. Another idea is that climatic stability permitted highly specialized species to evolve whose narrow niches created *ecological breathing space* that allowed many different specialized forms to live side by side. A third hypothesis is that stability, combined with geographical barriers and variations in factors like soils, allowed the evolution of large numbers of regionally localized specialists—endemic species—those unique to just one rainforest region such as the Atlantic Forest of Brazil. In fact, it is likely that several or even all of these factors have played a role, and active ecological research continues in surviving rainforest areas to try to find answers to these important questions.[72]

Despite the challenges of carrying out field studies in tropical moist forests (the technical term for rainforest used by scientists) a considerable amount has been learned about their structure and workings. Extensive research undertaken since the mid-twentieth century has carefully documented their remarkable physical structure, and has found that the shape, size and layering of vegetation is an important key to the coexistence of many thousands of species in relatively small areas. The first striking feature of this architecture is the great height of many rainforest trees which often reach 150 feet—roughly the height of a fifteen-story building—compared with 40–60 feet in many temperate forests. Tropical forest trees are not actually the tallest in the world, even though some species present in Southeast Asian rainforests do rise to well over 250 feet. The very tallest trees are found in quite localized temperate zone habitats: along North America's west coast where redwoods, Douglas firs and Sitka spruces can rise to 300 feet or more; and in southern Australia, where several species of eucalyptus also approach 300 feet. These very tall temperate forests, however, lack many of the vegetation elements that give rainforests their unique structural complexity.[73]

In many temperate forests, the basic physical structure of the community is relatively simple, with the trunks of the trees resembling a series of fairly regularly spaced vertical columns. In regions where coniferous species dominate, the tops of the trees may also have a striking uniformity of branches and needles; in broad-leaf forests, the tops of individual trees usually vary more in shape since growth increases on the side of the tree where there is more light. Temperate forests may also contain some woody shrubs or small, under-story trees, but that is generally the extent of their overall forest architecture. Tropical moist forests, on the other hand, incorporate the basic key elements of more northern forests, but add to them a whole spectrum of additional structures to create a vastly more complex living environment. Broad-leaved trees predominate in them—conifers are relatively rare in tropical lowland forests—but they can be of widely differing heights, and sometimes end up forming several layers, while most temperate forests have just a single layer. In many tropical forests, there is a significant layer of under-story trees, above which grows a canopy of full-sized species. And in tropical moist forests, there are also often scattered trees which rise up even higher than the canopy trees, sticking their branches up above the canopy to form scattered skyscraper trees called *emergents*. In addition to tropical rainforest trees' astounding diversity—in many areas nearly every tree is a different species from all those around it—such equatorial forests also contain two groups of plants that generally play minor roles, if any, in most temperate forests: vines or climbers, and air plants or *epiphytes*.[74]

Vines and climbers survive in one of two ways. Some grow up from the ground, using the stems and trunks of shrubs and trees for support until their leaves are high enough up to share in the canopy's brighter light. Others have seeds that germinate high up on a branch and send long, dangling roots ever downward until they plant themselves in the soil, after which a stem grows upward from the seed towards the light. Epiphytes, on the other hand, dispense entirely with roots anchored in the ground. Instead, their seeds germinate high up on tree branches and stems and then spend their lives as relatively small plants getting all their water from rain and mist, often storing some in little water tanks they form with their own leaves; their nutrients come from decaying tree bark, dust and animal droppings. Theirs is a spare and specialized way of life, although a very successful one. Without the enormous energy cost of growing a stem all the way up from the ground, their strategy nonetheless permits them to live high up in the well-illuminated canopy, or even in nearly full sunshine on the high branches of the tallest emergent trees. And grow they do. In one plot in Costa Rica sampled in the mid-1980s, an area of just 120 square yards (100 square meters) contained 59 different species of epiphytes, along with 30 kinds of vines and climbers, and 144 species of trees and shrubs. Altogether, this tiny area held 233 different kinds of higher plants plus 32 types of liverworts and mosses. And even this site is probably not the richest rainforest area in the world.[75]

In the western Amazon, near the border of northeastern Peru and southeastern Ecuador lies the Yasuni Rainforest, which is one of the most diverse habitats on earth. To grasp how richly it is stocked with life, consider its flora in relation to that of Great Britain. That large European island contains only about fifty kinds of native trees and shrubs in a total area of about 23,000,000 hectares. In a recent survey in the Yasuni forest, by comparison, a single area of only 25 hectares (sixty acres)—roughly one-millionth the area of Great Britain—was found to contain more than 1,100 different kinds of trees and shrubs.[76] Surveys of the two-and-a-half-square-mile Tiputini Biodiversity Station within Yasuni also found uniquely high richness of birds (550 species), reptiles and amphibians (247 species) and mammals (200 species) representing probably the richest land vertebrate fauna anywhere on earth.[77] Once all living species in the Yasuni plots are eventually counted—a daunting task due to the vast, little studied diversity of invertebrates and microbes that are found there, and the general neglect and under-funding of research on biological classification—hundreds of thousands of living species will almost certainly be identified within the park and surrounding areas of forest.[78]

The rainforest's treasure houses of life represent the most extensive set of solutions ever developed to the basic problems of survival on earth. Among them are a vast and unique set of structural, behavioral and biochemical responses to nearly every kind of problem that living things face—from defense against predators and parasites, to the need for disguise and camouflage, to resistance to infection by bacteria, viruses, parasites and fungi. Many researchers are convinced that large numbers of yet to be discovered medicines and unique biochemical molecules will eventually be found hidden within the rainforest's unparalleled concentration of life. Such habitats are also critically important in helping to maintain long-term global ecological stability and evolutionary potential in the face of unprecedented human environmental impact. They do this in many different ways, most obviously by storing enormous amounts of carbon which they chemically lock up in living vegetation.

Systematic scientific exploration of living rainforest creatures began only in the nineteenth century, and progressed very slowly until the 1960s. Since then, more scientists have carried out additional studies, but the sum total of what there is to learn is still so great that only a small sample of rainforest mysteries have so far been deciphered. Yet tragically, rainforests are being so rapidly destroyed or degraded in many places that most will have disappeared forever before scientists are able to even document what is being lost. Of all earth's conservation priorities, preserving extensive, undisturbed areas of intact rainforest in as many locations as possible must be at the top of the list, for it is there that the richness of life is most heavily concentrated.

People and Tropical Forests

It is likely that people have inhabited tropical moist forests, at least in small numbers, for a very long time, although they are environments in which an abundance of insects, microbes and parasites ensure that considerable skill and knowledge are required for survival. As farming spread among different human groups, some rainforest dwellers also began to grow some of their own food, although we may never gain a complete understanding of the actual extent of indigenous agriculture in lowland rainforests in the days before advancing sailing ship technology allowed global exploration and trade. Along the major rivers in the Amazon Basin, for example, there is evidence of more permanent and extensive farm fields, perhaps fertilized by intentionally buried charcoal and animal bones.[79] In many places, however, only shifting cultivation seems to have been practiced, in which the trees in small patches of forest were first cut and then burned, followed by cultivation of the soil for one to three years. At the end of that time, people abandoned the cleared patch allowing it to regenerate as forest, while they moved to a new location and proceeded to clear another small patch of forest. Unlike modern land clearing in places like Brazil and Indonesia, this slash and burn subsistence was done on patches that were small enough for the natural plant community to regenerate over time because creatures remaining in the nearby surviving, intact rainforest could recolonize the burned patches once they were abandoned.

What we do know is that throughout much of the nineteenth century, as modern botanists, zoologists and geographers first began to systematically explore these living communities, large areas of intact rainforest, teeming with unparalleled living richness, were still present in equatorial Africa, South America and Southeast Asia, as well as in central America, Indo-China, New Guinea, Madagascar, and a few other locales.[80] By the twentieth century, however, this had all begun to change, as a wide range of advances in medical, mechanical and engineering technologies—improved sanitation and disease control, enhanced road building, gasoline and diesel powered boats, trucks, tractors, bulldozers and chain saws—fundamentally shifted the potential for human settlement and impact in these regions. The result was a population explosion which, in turn, led to widespread destruction and degradation of vast areas of rich tropical forests.[81] Two examples illustrate this, one from southeast Asia and one from South America.

Indonesia is a medium-sized country by area, sprawling across 13,000 islands in tropical southeast Asia. In 1800, its total population was just eighteen million, and nearly all of its rich, tropical forests were still intact. During the 1850s, for example, the English biologist Alfred Russell Wallace—the co-discoverer with Charles Darwin of the process of evolution—spent eight years traveling in the region making unique observations and collections of its remarkable fauna and flora. By 1900, however, Indonesia's population had more than doubled to about 40 million, creating pressure to convert more land to agriculture. Yet even then, its greatest population explosion still lay ahead of it. Over the course of the twentieth century, its population grew more than two and a half times as fast as it had during the nineteenth century, bringing the national census in 2000 to more than 211 million, a level more than ten times as high as in 1800 (an increase of over 1,000 percent in just 200 years). By 2016, it had already exceeded 260 million.[82]

According to a report issued by the World Resources Institute (a non-profit environmental advocacy group), by the year 2000 Indonesia had lost nearly three quarters (72%) of its original forests, with 40 percent having been cut down in just the five decades from 1950 to 2000.[83] And if this were not damaging enough, the rate of cutting and burning was continuing to accelerate in the twenty-first century. Between 2000 and 2012, more than six million hectares were lost, an area nearly equal in size to the entire country of Sri Lanka, with predictions that several Indonesian islands may lose essentially all of their original forest early in the twenty-first century.[84] This kind of wholesale forest destruction entails not only disastrous losses of biodiversity, but also generates massive amounts of soil erosion along with air and water pollution.[85] Enormous amounts of carbon dioxide and methane are released when large fires are set to reduce vegetation to ashes, making a significant contribution to global warming.

Our other example comes from the New World, where the Atlantic Forest of Brazil (Mata Atlantica) used to stretch more than 1,500 miles from the northeastern corner of Brazil down to Paraguay and northeastern Argentina, covering about 1,200,000 square kilometers—an area roughly the size of South Africa. Long occupation by indigenous people seems to have left the habitat largely intact until European settlement initiated permanent forest clearing in the sixteenth century. And even as late as the mid-nineteenth century, when Charles Darwin visited during his voyage on HMS *Beagle*, the Atlantic Forest still contained the majority of its original biological riches. So full of life was the original region, that it contained almost as many species as the whole Amazon, an area several times larger. Even today, it is estimated that the surviving fragments contain 20,000 different species of plants—fully 8 percent of the world's flora—and that 8,000 of them grow nowhere else on earth. On just two and a half acres (one hectare), for example, researchers were able to identify 458 different kinds of trees, more than in all of eastern North America. In addition, the Atlantic Forest supports 2,200 different kinds of vertebrates—mammals, birds, amphibians and reptiles—5 percent of all known vertebrates on earth, including 934 kinds of birds, and dozens of species of primates.[86]

From the mid-nineteenth century on, the Atlantic Forest was increasingly cut for timber, charcoal, cattle ranches, sugar cane fields and coffee plantations. Where it was not cleared for either agriculture or urbanization—the mega-cities of Rio de Janeiro and São Paulo both lie within its natural boundaries—ancient, native forest was often replaced by plantings of exotic, fast-growing trees, particularly species of Australian Eucalyptus. Such plantations commonly contained one or just a handful of species, while the natural forests they displaced were among the most diverse on earth. By the early twenty-first century, nearly 92 percent of the Atlantic Forest had been cut down, leaving behind small and disconnected surviving patches, many missing important native species, and mostly inhabited by small, isolated natural populations. Under such conditions, the long-term survival of thousands of its species remains gravely in doubt. Yet even after such terrible losses, the surviving 8 percent of the Atlantic Forest still represents one of the greatest treasure houses of life on earth. Steps are being taken to protect what remains, but the story of the Mata Atlantica provides a powerful cautionary tale of how rapidly the earth's survival wisdom can be destroyed if people fail to grasp and appreciate its true nature.

Conclusion

The human species is absolutely remarkable. Beginning hundreds of thousands of years ago with simple stone tools and tiny populations, we have risen through one technological advance after another to the point where we have overpowered much of the rest of the living world.[87] The rise of modern science in the last few hundred years has dramatically accelerated the pace and scope of human invention. We have displaced natural communities on the earth's most fertile lands on a scale so grand, that if assembled in one piece, crop fields alone would cover an area nearly the size of South America. Grazing land covers an area nearly twice as large, and cities, towns and forestry plantations further increase the area of natural communities damaged or destroyed.

In the short run, this may not seem to matter much—the vast habitat destruction that has occurred on every continent except Antarctica, and which includes many rich tropical areas like the Indonesian rainforest and the Atlantic Forest, has not so far brought the world to an end; and beef, coffee, sugar, rice, palm oil and tropical fruits are all important, economically valuable parts of the human diet and economy. But after half a billion years of life on dry land, and at least two million years of human life, the popular notion that we should do whatever we want with the earth's resources because what may happen a hundred or a thousand years from now is unimportant or irrelevant now seems reckless and even criminally short-sighted.

The stark fact confronting us as we face the future is that we have already damaged or destroyed more natural land communities than still survive intact. Yet even in the face of such incalculable losses, a combination of very low levels of scientific education and literacy, large family sizes in some countries and among certain religious communities, and endless economic desires, mean that land *clearing* and the destruction of natural communities is still continuing at a fast pace. Truly, this is one of the world's most pressing, if least discussed, problems.

Surviving intact habitats, especially tropical forests, are crucial to global ecology, and play critical roles in the worldwide flows of nutrients and energy. In addition to their beauty, structural intricacy and complex behavior, the creatures that survive in them contain unique stores of genetic knowledge that allowed them to survive on earth for millions of years. So far, we have not yet learned enough to decipher most of that wisdom. But in coming centuries scientists will increasingly be able to unravel more and more of these natural survival solutions. These can provide both better models to guide our own long-term success and large numbers of important, new and sustainable technologies based on naturally evolved substances, processes and relationships. If, however, we continue to destroy the vast majority of earth's remaining forests and other intact habitats, many of the species and unique interactions they contain will simply vanish. And if this happens, our descendants will forever be denied much of the vast treasure trove of wisdom about long-term survival that wild creatures have so painstakingly accumulated over millions of generations.

Notes

1   R. G. Allaby et al., 2008, The genetic expectations of a protracted model for the origins of domesticated crops. *PNAS* 105(37): 13,982–13,986.

2   Hunter-Gatherer.org, Facts and theories about hunter-gatherers, http://hunter-gatherers.org/facts-and-theories.html; Wikipedia, Hunter-gatherers, https://en.wikipedia.org/wiki/Hunter-gatherer.

3   US Census Bureau, World populations, www.census.gov/population/international/data/worldpop/table_history.php; World Population Growth; World Bank, 2004, www.worldbank.org/depweb/beyond/beyondco/beg_03.pdf.

4   H. Levin, 2006, *The Earth Through Time*, 8th edition, Wiley.

5   Ibid.

6   United States Census Bureau, Historical estimates of world population, www.census.gov/population/international/data/worldpop/table_history.php.

7   New World Encyclopedia, Human evolution, www.newworldencyclopedia.org/entry/Human_evolution#Neanderthals.

8   A. Sharifi et al., 2015, Abrupt climate variability since the last deglaciation based on a high-resolution, multi-proxy peat record from NW Iran: the hand that rocked the cradle of civilization, *Quaternary Science Reviews* 123: 215–230.

9   About.com Archaeology, Plant domestication table of dates and places, http://archaeology.about.com/od/domestications/a/plant_domestic.htm; R. M. Bourke, History of agriculture in Papua New Guinea, http://press.anu.edu.au/wp-content/uploads/2011/05/history.pdf.

10 J. Cornell, 2011, Slash and burn, Encyclopedia of Earth, www.eoearth.org/view/article/156045.

11 About.com Archaeology, Animal domestication table of dates and places, http://archaeology.about.com/od/dterms/a/domestication.htm.

12 C. K. Pearse, 1971, Grazing in the Middle East: past, present and future. *Journal of Rangeland Management*: 24(3): 13–16.

13 USDA, Hubbard Brook ecosystem study, www.hubbardbrook.org.

14 Ibid.

15 D. J. Fairbanks and W. R. Andersen, 1999, *Genetics: The Continuity of Life*, Brooks/Cole.

16 Ibid.

17 The Florida Panther Society, www.panthersociety.org/moreinfo.html; C. F. Facemire et al., 1995, Reproductive impairment in the Florida panther: nature or nurture? *Environmental Health Perspectives* 103(Suppl. 4): 79–86.

18 M. Elvin, 2004, *The Retreat of the Elephants: An Environmental History of China*, Yale University Press.

19 Wikipedia, List of famines in China, https://en.wikipedia.org/wiki/List_of_famines_in_China.

20 Israel Science and Technology, List of periodic table elements sorted by abundance in Earth's crust, www.science.co.il/PTelements.asp?s=Earth.

21 John Deere, John Deere's plow, www.deere.com/en_US/corporate/our_company/about_us/history/john_deere_plow/john_deere_plow.page; ChinaCulture.org, Iron plow, www.chinaculture.org/gb/en_madeinchina/2005-04/30/content_68358.htm.

22 Oregon State University, Erosion, http://people.oregonstate.edu/~muirp/erosion.htm.

23 N. Rodgers, 2008, *Roman Empire*, Metro Books.

24 Wikipedia, Horse collar, https://en.wikipedia.org/wiki/Horse_collar; Equisearch, The history of horseshoes, www.equisearch.com/article/history-of-horseshoes-17802.

25 Encyclopedia Britannica, Two field system, www.britannica.com/topic/two-field-system.

26 The Open Door Web Site, The four field system, http://saburchill.com/history/chapters/IR/003f.html.

27 The Open Door Web Site, The seed drill, http://saburchill.com/history/chapters/IR/004f.html; Listverse, 10 great Chinese inventions, http://listverse.com/2009/04/18/10-great-ancient-chinese-inventions/

28 About.com, Cyrus McCormick, inventor of the mechanical reaper, http://inventors.about.com/od/famousinventions/fl/Cyrus-McCormick-Inventor-of-the-Mechanical-Reaper.htm.

29 US Census Bureau, World population, www.census.gov/population/international/data/worldpop/table_history.php.

30 But see M. Elvin, 2004, *The Retreat of the Elephants: An Environmental History of China.* Yale University Press

31 Population Reference Bureau, Human population: population growth, www.prb.org/Publications/Lesson-Plans/HumanPopulation/PopulationGrowth.aspx.

32 D. J. Fairbanks and W. R. Andersen, 1999, *Genetics: The Continuity of Life,* Brooks/Cole.

33 Wikipedia, Nazareno Strampelli, https://en.wikipedia.org/wiki/Nazareno_Strampelli.

34 Information for Forest Management, The history of the chain saw, www.waldwissen.net/lernen/forstgeschichte/wsl_geschichte_motorsaege/index_EN.

35 US Census Bureau, World population, www.census.gov/population/international/data/worldpop/table_history.php.

36 Saritha Pujari, Decline in death rate—a major cause of human population in India, www.yourarticlelibrary.com/essay/decline-in-death-rate-a-major-cause-of-human-population-in-india/26917.

37 US Department of Health and Human Services, The great pandemic, www.flu.gov/pandemic/history/1918.

38 S. N. Shchelkunov, 2011, Emergence and reemergence of smallpox: the need for development of a new generation smallpox vaccine, *Vaccine* 29(Suppl. 4): D49–D53; S. Riedel, 2005, Edward Jenner and the history of smallpox and vaccination, *Baylor University Medical Center Proceedings* 18(1): 21–25.

39 Wikipedia, Smallpox, https://en.wikipedia.org/wiki/Smallpox.

40 Wikipedia, History of smallpox, https://en.wikipedia.org/wiki/History_of_smallpox.

41 T. Mestrovic, Smallpox history, *News Medical,* www.news-medical.net/health/Smallpox-History.aspx.

42 Wikipedia, History of smallpox, https://en.wikipedia.org/wiki/History_of_smallpox; S. Riedel, 2005, Edward Jenner and the history of smallpox and vaccination, *Baylor University Medical Center Proceedings* 18(1): 21–25.

43 PBS, World Health Organization declares smallpox eradicated, www.pbs.org/wgbh/aso/databank/entries/dm79sp.html.

44 Wikipedia, Pasteurization, https://en.wikipedia.org/wiki/Pasteurization.

45 D. Pitt and J. M. Aubin, 2012, Joseph Lister: father of modern surgery, *Canadian Journal of Surgery* 55(5): E8–E9.

46 Wikipedia, Anthrax, https://en.wikipedia.org/wiki/Anthrax.

47 Encyclopedia Britannica, Louis Pasteur, Vaccine development, www.britannica.com/biography/Louis-Pasteur/Vaccine-development.

48 Haffkine Institute, 2015, Waldemar Mordecai Haffkine, www.haffkineinstitute.org/waldemar.htm.

49 World Health Organization, Cholera, www.who.int/mediacentre/factsheets/fs107/en.

50 UCLA Department of Epidemiology, John Snow, www.ph.ucla.edu/epi/snow.html; J. Snow, 1855, On the Mode of Communication of Cholera, www.ph.ucla.edu/epi/snow/snowbook.html.

51 Wikipedia, London sewerage system, https://en.wikipedia.org/wiki/London_sewerage _system.

52 Ibid.

53 123helpme!com, History of rabies, www.123helpme.com/history-of-rabies-view.asp?id=154021.

54 M. Zaitlin, 1998, The discovery of the causal agent of the tobacco mosaic disease, in S. D. Kung and S. F. Yang (eds.), *Discoveries in Plant Biology*, World Publishing Company, pp. 105–110.

55 H-W., Ackermann, 2011, The first phage electron micrographs, *Bacteriophage* 1(4): 225–227, www.ncbi.nlm.nih.gov/pmc/articles/PMC3448108.

56 US Census, World population, www.census.gov/population/international/data/worldpop/table_history.php.

57 Wikipedia, Age of discovery, https://en.wikipedia.org/wiki/Age_of_Discovery.

58 Wikipedia, Hispaniola, https://en.wikipedia.org/wiki/Hispaniola.

59 See, for example, C. Mann, 2005, *1491: New Revelations of the Americas Before Columbus*, Random House; and C. Mann, 2011, *1493: Uncovering the New World Columbus Created*, Random House.

60 University of California Museum of Paleontology, Carl Linnaeus, www.ucmp.berkeley.edu/history/linnaeus.html.

61 Wikipedia, Evolution, https://en.wikipedia.org/wiki/Evolution.

62 Momtastic, Incoming! The world's 10 worst invasive species, http://webecoist.momtastic.com/2009/12/15/incoming-the-worlds-10-worst-invasive-species; Wikipedia, List of globally invasive species, https://en.wikipedia.org/wiki/List_of_globally_invasive_species.

63 R. P. Duncan et al., 2013, Magnitude and variation of prehistoric bird extinctions in the Pacific, *PNAS* 110(16): 6436–6441, www.pnas.org/content/110/16/6436.

64 M. E. Dorcas et al., 2012, Severe mammal declines coincide with proliferation of invasive Burmese pythons in Everglades National Park, *PNAS* 109: 2418–2422.

65 N. Ramankutty and J. A. Foley, 1999, Estimating historical changes in global land cover 1700–1992, *Global Biogeochemical Cycles* 13(4): 997–1027.

66 C. Monfreda et al., 2008, Farming the planet: 1. Geographic distribution of global agricultural lands in the year 2000, *Global Biogeochemical Cycles* 22(1): 1–19; World Almanac, 2006, *World Almanac*, World Almanac Books.

67 Wikipedia, List of sovereign states and dependencies by area, https://en.wikipedia.org/wiki/List_of_sovereign_states_and_dependencies_by_area; World Almanac, 2006, *World Almanac*, World Almanac Books.

68 UN Food and Agriculture Organization, 2014, Global data on forest plantations resources, www.fao.org/docrep/004/Y2316E/y2316e0b.htm.

69 C. Monfreda et al., 2008, Farming the planet: 1. Geographic distribution of global agricultural lands in the year 2000, *Global Biogeochemical Cycles* 22(1): 1–19; Wikipedia, World population, https://en.wikipedia.org/wiki/World_population.

70 G. T. Miller, Jr., 2006, *Environmental Science*, 11th edition, Brooks-Cole.

71 T. C. Whitmore, 1992, *An Introduction to Tropical Rain Forests*, Oxford University Press; D. Holing and S. Forbes (eds.), 2001, *Rain Forests*, Discovery Communications; E. P. Odum and G. W. Barrett, 2005, *Fundamentals of Ecology*, Brooks/Cole.

72 J. Terborgh et al., Duke University Biology Department, 2002, Maintenance of tree diversity in tropical forests, http://people.duke.edu/~ncp/Homepage_of_Nigel_Pitman,_Duke_University_NSOE_Center_for_Tropical_Conservation,_Amazonian_research_and_conservation/Publications_files/terborgh2002.pdf.

73 T. C. Whitmore, 1992, *An Introduction to Tropical Rain Forests*, Oxford University Press.

74 Ibid.

75 T. C. Whitmore et al., 1985, Total species count in a Costa Rican tropical rain forest, *Journal of Tropical Ecology* 1: 375–378.

76 N. C. A. Pitman et al., 2002, A comparison of tree species diversity in two upper Amazonian forests, *Ecology* 83(11): 3210–3224.

77 University of Texas at Austin, January 19, 2010, Scientists identify Ecuador's Yasuni National Park as one of most diverse places on earth, www.utexas.edu/news/2010/01/19/ecuador_biodiversity.

78 M. S. Bass et al., 2010, Global conservation significance of Ecuador's Yasuni National Park, *PLOS One* 5(1): e8767, www.plosone.org/article/info%3Adoi%2F10.1371%2F journal.pone.0008767.

79 Terra preta de indio, 2014, Cornell University Department of Crop and Soil Sciences. www.css.cornell.edu/faculty/lehmann/research/terra%20preta/terrapretamain.html.

80 C. Darwin, 1962, *The Voyage of the Beagle*, Doubleday & Co.; A. R. Wallace, 1962, *The Malay Archipelago*, Dover Publications.

81 T. C. Whitmore, 1992, *An Introduction to Tropical Rain Forests*, Oxford University Press.

82 Worldometer, Indonesia population, www.worldometers.info/world-population/indonesia-population.

83 World Resources Institute, 2016, *Forest Cover, 1990–2005*, Indonesia, www.wri.org/resources/charts-graphs/forest-cover-1990-2005-indonesia.

84 B. A. Margono, et al., 2012, Primary forest cover loss in Indonesia over 2000–2012, *Nature Climate Change* 4: 730–735.

85 J. Vidal, 2014, Indonesia's forest fires feed "brown cloud" of pollution choking Asia's cities, *The Observer* (March 22), www.theguardian.com/environment/2014/mar/23/indonesia-forest-fires-pollution-asia.

86 M. C. Ribeiro, 2009, The Brazilian Atlantic Forest: how much is left, and how is the remaining forest distributed? Implications for conservation, *Biological Conservation* 142: 1141–1153.

87 E. O. Wilson, 2013, *The Social Conquest of Earth*, W. W. Norton.

4

MANUFACTURED CHEMICALS

Introduction

Before permanent agriculture, harmful human impacts on nature were mostly limited to the over-hunting of animals and the burning of particular habitats for game management. By 10,000 years ago, however, this all started to change as farmers began to replace existing plant communities with cultivated fields—a process that has continued ever since in response to continually expanding human population. Neither of these impacts on living communities is particularly hard to understand: if we eat too many animals, there will be less of them, or none at all, and their natural predators may also disappear; if we keep chopping down and plowing up forests and grasslands, very few of their original inhabitants will still be present as time goes on. Humanity's third major impact on nature, however, is much less obvious, and it involves forces that are largely invisible. Its damaging events often occur hidden away inside large factories or within the tissues of living things, and understanding them requires some familiarity with science and engineering. These are the impacts that result from the creation and use of a wide range of manufactured chemicals.

This chapter and the one which follows aim to examine some of the effects of manufactured synthetic and natural chemicals on the living world and on human health. To fully treat this topic would require a work of encyclopedic length; here we will cover just a few examples, chosen to represent some of the most serious threats chemicals pose to the survival of life. To begin, we will go back in time and look at the ways that people have found and utilized new substances since they began to extract metals more than 8,000 years ago. Use of new materials has always been fundamental to the advance of technology. But it took many, many centuries before slowly acquired hands-on experience and the speculations of alchemists finally gave way to chemical and material science and the modern chemical industry. And once that transition occurred in the nineteenth century, the pace of discovery and release of manufactured chemicals increased dramatically.

In today's industrialized world we are confronted with a chemical environment completely unlike the one our stone age ancestors inhabited, one containing tens of thousands of manufactured substances, many of which are entirely new to nature. Large amounts of such materials began to be produced in the second half of the nineteenth century, and their variety and quantity has continued to increase throughout the twentieth century and up to the present. The twentieth century, however, also saw unprecedented advances in medicine and biology, including the creation of the new scientific fields of ecology, public health and environmental toxicology. Together, these have documented the complex and far-reaching damage caused by widespread use of some manufactured substances. In a few cases, partial or complete bans on particular compounds were imposed by individual countries or by international agreement. Yet the range of synthetic chemicals is so great, and the workings of human, animal and plant biology so complex, that it is nearly impossible to thoroughly test the entire range of synthetic chemicals for possible harm. For this reason, the overwhelming majority of these new substances still remain in use. Many are probably harmless, or may cause only minor damage, while others likely represent unrecognized long-term threats either on their own or in combination with other chemicals that are also present in the environment.

Clay, Metal, and Fire

After stone tools, fire was humanity's second great technology, one that our ancestors began using as long as 300,000 years ago.[1] Fire is so commonplace to us that we have to make a real effort to put aside our familiarity with it in order to remind ourselves how very unusual and remarkable it is. Even our very ancient forbears must have wondered what unseen processes were going on in the flames they so carefully tended for cooking, lighting, warmth and defense against predators. Thick, heavy tree branches would go into the fire and soon burst into flames, almost as if they were alive. Before long, however, they literally began to disappear, and after only a few hours, the wood was all gone, leaving behind only a tiny pile of lightweight coals and ash. Fire was like nothing else that our ancestors knew or encountered. It was unique not only because it was so useful, but also because of its odd habit of transforming things—making flames, light and heat appear, while at the same time making wood and other flammable materials vanish.

Over millennia of fire use, people also noticed that it could transform materials other than wood, although usually without making them disappear. Clay could be baked into ceramics without being consumed, and certain kinds of rock, when accidentally heated, produced small lumps of shiny new substances like lead or tin, metals that could be *smelted* in a hot campfire. Baked clay came first, with ceramic figurines being made as long ago as 27,000 years.[2] The earliest fired clay pots discovered so far were made 18,000 years ago in south China.[3] By 8,500 years ago, there is clear evidence of the use of extracted metals, notably in the form of lead beads found in the excavated remains of the village of Catalhoyuk in modern Turkey. Metallic lead is not found in nature, and Catalhoyuk's inhabitants may have been the first people to intentionally extract it from rock.[4] Tin and lead are the easiest of all metals to produce in this way as both melt at temperatures easily reached in a hot wood fire. By 5,000 years ago in Serbia, members of the Vinča culture had managed to solve the far more difficult problem of creating fires that were hot enough to extract copper from rocks, a process requiring temperatures close to 2,000°F. To do this, they learned to burn charcoal rather than wood and to blast extra air through the fire using simple bellows. Initially, the Vinča seem to have used copper primarily for decorative objects, but later they also employed it in axes and other tools.[5]

The smelting of metals clearly represented an enormous advance in human technology, but it also had a downside. Given the history of earth's formation and the varied geological processes that continue to this day, many rocks contained complex mixtures of elements. When they are heated to release pure metals, they also often release hazardous substances. For this reason, the early production of metals created the first man-made chemical pollution and also the first chemical occupational hazard for the workers who extracted them. Some authors even think that the legendary *lameness* of the metal-working gods in different European traditions—Hephaistos in Greece, Vulcan in Rome, Wayland in England—reflects the frequent occupational poisoning of early metalworkers by lead, arsenic and other toxic substances that are often present in ore rocks.[6] (Some of the underlying mechanisms that cause harm to people and other living things following exposure to these metals are explored in the next chapter.)

Toxic metal poisoning in ancient times seems to have reached its greatest extent in Rome, a civilization which made unprecedented use of lead in water pipes, in vessels used to store and prepare food and drink, and in cosmetics. Indeed, some scholars believe that Rome's cultural, economic and political decline was partly caused by an epidemic of lead poisoning that most heavily impacted the upper classes. So widespread was the smelting of lead in Roman times that clear evidence of a peak in European-wide atmospheric lead pollution is detectable in sediments dating from that era. During the same period, the Romans also made extensive use of coal—particularly in Britain where it was abundant—and this also led to significant dispersal into the environment of toxic mercury, since it is commonly found as an impurity in coal beds. Once the Roman Empire had collapsed, lead was used much less widely in Europe, and lead poisoning seems to have declined, although not completely ended.[7]

Alchemy

Early observations of campfires and the ability of flames to transform certain substances led many people to be intensely curious about metals—attractive materials that were often shiny, heavy and very different from most other familiar objects. Lead and tin were probably both discovered after ore-bearing rocks were accidentally heated in campfires—neither is particularly common in the earth's crust. Lead is too soft to be of much use for tools, but tin can be melted together with copper—a metal that occasionally occurred naturally in pure form—to make the alloy bronze, a metal hard enough to use for knives and other tools. In addition to copper, gold and silver also sometimes occur naturally as pure metal, although often they are combined with one another in rocks or otherwise require refining. Silver is hundreds of times rarer than copper, and gold is many times rarer than silver. Both are striking, shiny, shapeable and resistant to discoloration. Even though they are too soft for tools, their attractiveness and scarcity soon led to their being considered *precious* metals; even 6,000 years ago, decorative gold objects were being made in Eastern Europe, and by 3,500 years ago gold had also begun to serve as a *currency* of international trade. It was simultaneously a highly valuable possession and a substance of great beauty and ceremonial importance.[8]

As gold became more and more valuable, people focused on acquiring more of it, and some—probably inspired by the transforming powers of fire—began to search for ways to change more common metals into gold or silver. Today we know that this is not actually possible except through the extreme and prohibitively expensive techniques of nuclear reactions which produce gold that is radioactive. But the desire for gold was so strong that for several thousand years, many people devoted tremendous time and energy to efforts designed to create it. These were the *alchemists* who were also often deeply interested in the study of astrology, religion, and what today would be called *magic*.

Alchemy's history is difficult to unravel. Many practitioners intentionally kept their ideas and methods secret or shared them only within a narrow circle, and early records were often lost or destroyed in the course of wars, conquests and religious conflicts. For all these reasons, early alchemy does not seem to have made any significant progress in creating practical chemistry. Instead, limited practical advances continued to be made by artisans and skilled workers in ceramics, metals and other materials as they developed their crafts by trial and error. As with alchemy, however, the successes of artisans in individual countries were often closely guarded secrets. Their benefits were not made available to a wider community, and they were easily lost in the event of social upheaval.

The beginnings of chemical progress beyond these limited confines began in the eighth century in the Middle East. It was led by a series of brilliant alchemists beginning with Jabir ibn Hayyan, who came to be known in the West as Geber. Ibn Hayyan's view of nature was pre-scientific and he held a number of beliefs that had little basis in fact, but he was also a brilliant and creative researcher who still managed to make some real progress in practical chemical technology. Affirming the fundamental importance of doing experiments, he was able to demonstrate processes like crystallization and distillation, and to invent some of the earliest workable vessels for chemical processes and experiments. He also developed more rigorous methods of carrying out reactions and recording his results. Along with other alchemists in his tradition, he was responsible for the discovery of a number of very important chemicals including acetic, hydrochloric, nitric and sulfuric acids.

After ibn Hayyan, a successful Middle Eastern school of alchemy survived for several hundred years, in spite of setbacks during the rule of those Caliphs who were religious fundamentalists. In 1258, however, this tradition came to a sudden end when Mongol armies captured and sacked Baghdad destroying researchers' laboratories and records along with the city's great libraries of manuscripts. After the fall of Baghdad, alchemy's doctrines and beliefs did not simply disappear, but instead continued along for hundreds of years without either realizing their stated goals or adding much to practical chemical technology.

The Science of Chemistry

The emergence of scientific chemistry from alchemy occurred gradually over the course of the seventeenth and eighteenth centuries, inspired by advances in astronomy and physics and by a new emphasis on empirical inquiry. The spirit underlying this new approach was laid out by Francis Bacon, who in 1605 described what we would now call the scientific method in a work called *Of the Proficience and Advancement of Learning, Divine and Human.*

One important figure who helped lay the foundations for scientific chemistry was the English scientist Robert Boyle (1627–1691), who is remembered today for his discoveries in both chemistry and physics. In keeping with long-standing alchemical traditions, Boyle still believed that there might be a way to transform ordinary metals into gold, but following Bacon's ideas, he was also passionately committed to establishing knowledge only through sensory experience and experiments. He spent years doing research, all the while keeping careful, quantitative records of his findings. He also made important contributions to theoretical and practical chemistry, summarizing his findings in a book called *The Scyptical Chymist*, published in 1661. Among other advances, Boyle clearly distinguished pure chemical compounds from mixtures and proposed that something like chemical elements existed. He also made advances in what he termed *chemical analysis*. He was a tireless researcher who kept extremely precise records and also readily shared his findings with other scientists.[9]

Following *The Scyptical Chymist*, it seems to have taken the better part of a century for progress in scientific chemistry to speed up. By the mid-eighteenth century, however, important discoveries started to be made every decade, with notable progress being made in the study of gases. In 1754, for example, the Scotsman Joseph Black discovered a gas he referred to as "fixed air"—today's carbon dioxide; in 1766 the Englishman Henry Cavendish isolated hydrogen gas; and in the early 1770s, the Swedish-German researcher Carl Wilhelm Scheele and the Englishman Joseph Priestley independently discovered oxygen. Against this background of progress, chemistry was soon poised to emerge as a real science, and the figure most responsible for its crossing that boundary was the French chemist Antoine Lavoisier.

In the 1770s, Lavoisier set up a state-of-the-art chemical laboratory in Paris, where—equipped with the best apparatus available—he consistently made very precise measurements of weights, volumes, temperatures and other reaction conditions. He kept clear records and even repeated his own experiments to verify their results. Unlike the alchemists, he promptly and publicly described and explained his experiments and their outcomes. In one notable breakthrough, he demonstrated again and again that *the weight of the products of any chemical reaction equaled the sum of the weights of the original substances that had been reacted*, a finding that came to be known as the law of *conservation of mass*. This turned out to be among the most basic of all chemical principles and one that proved essential to all further progress. In addition, Lavoisier collaborated with several other French chemists on a complete overhaul of the terms and definitions used in chemistry, in 1787 he co-authored a revolutionary work called *Method of Chemical Nomenclature*. Two years later he created what was essentially the first modern textbook of chemistry, the *Elementary Treatise on Chemistry*.

Following Lavoisier, chemistry enjoyed a remarkable period of excitement and discovery in the early decades of the nineteenth century, although much of the progress was made despite a still quite limited understanding of chemical elements, molecular structure and the actual mechanisms going on in reactions. Advances pioneered by early chemists like Lavoisier and the Englishman John Dalton now allowed laboratory researchers and factory chemists to try out new reactions, make new compounds, and discover different ways of creating important products and intermediaries.[10]

Early Chemical Industries—Acids, Alkali and Bleach

Over many thousands of years prior to the scientific revolution, craftspeople and inventors had gradually developed many practical material technologies in order to produce things like pottery and metals, textiles, tools, building materials, furniture, glass, soap, perfume, musical instruments and works of art. In many cases, the techniques and formulas involved were utilized only within single regions or kingdoms and were also kept secret. And like many early Middle Eastern chemical techniques, they were often lost over time due to wars and other upheavals. Ancient Rome, for example, produced some of the finest, most durable building concrete ever made—some of it is still structurally sound and aesthetically pleasing after 2,000 years. They did it with a special blend of ingredients that included lime and volcanic ash. But the formula was lost, and it was only in the twenty-first century that researchers managed to work out the details of their recipe. Now it is hoped that this discovery will lead to much more durable modern concrete, especially where it must be exposed to salt water.[11]

In Europe, by the eighteenth century, practical chemical technologies used in manufacturing had also begun to change. Modern chemistry was not yet well-developed but had begun to make progress; manufacturing, trade and population were steadily increasing; and new mechanical and power technologies were encouraging factories to grow larger. The result of all this was significantly increased chemical production particularly of mineral-based *inorganic chemicals*, those lacking carbon–hydrogen bonds. Prominent among these were acids, alkalis and a handful of other materials including lime, salts, nitrates and phosphates—substances widely used in the manufacture of textiles, glass, soap, and metal products. As is common with manufactured products, increased output tended to drive down prices, further stimulating demand, while lower prices also led to increased production of finished goods, reducing their price, making them available to more people, and allowing material standards of living to rise, at least for the middle and upper classes.

Acids were one important category of eighteenth-century industrial chemicals. They were used not only for the bleaching and dyeing of cloth, but also as a reagent for making other chemicals. Since ancient times, sulfuric acid had been made by a difficult, expensive process in which iron disulfide (pyrite) was broken down by heating in several steps to yield iron oxide and sulfur trioxide, with the sulfur trioxide then being dissolved in water to yield the acid. In the mid-1600s, however, a German alchemist and chemist named Johann Glauber developed a way to make dilute sulfuric acid by burning saltpeter (potassium nitrate) together with sulfur in the presence of steam. By the 1730s, the Englishman Joshua Ward had adapted this process to commercial production. Other manufacturers then worked on ways to concentrate sulfuric acid by distilling it, and by the 1740s John Roebuck had worked out how to make it in large vats lined with lead, a development that greatly increased production and lowered costs. By the 1760s mass production had also begun in France, and by 1800 the commercial price of sulfuric acid had dropped by 90 percent compared to the 1740s. Today, sulfuric acid remains one of industry's most important materials, with more than 150 million tons produced every year; it is used in batteries, mineral extraction from ores, fertilizer manufacture and chemical syntheses. Sulfuric acid is also a hazardous substance which produces intense heat when combined with water and has the ability to cause serious burns.[12]

Besides acids, the eighteenth-century chemical industry also produced ever-increasing amounts of alkalis, strong substances like sodium carbonate or sodium hydroxide (lye) which are the chemical opposite of acids. In our homes today, dilute alkalis are familiar components of many cleaning products including almost all soaps. Eighteenth-century industry required large amounts of strong alkalis, often called *soda* (especially sodium carbonate), for making glass, soap and—after the 1790s—bleach. In fact the importance of chemical production to the industrial revolution can be seen in the changes in the way that cloth was bleached before and after the middle of the eighteenth century. As late as the 1740s, the Netherlands still led other European countries in traditional methods of bleaching, with an industry centered around the town of Haarlem, not too far away from well-established flax-growing and linen weaving centers in Belgium. Traditionally, bleaching was carried out in large fields where newly woven cloth was exposed to the sun for weeks on end, while also repeatedly being treated with lye (alkali from plant ashes) and buttermilk (a weak acid). This whole process could take several months, required extensive space and considerable labor, and produced bleached cloth that was quite expensive. By the 1750s, however, new sources of dilute sulfuric acid began to be used to bleach the cloth in place of lye, buttermilk and sunlight, and this reduced the whitening time to less than one day. By the 1790s the first manufactured chlorine bleach had been developed in France, a product that allowed low-cost bleaching of cloth to be completed in a matter of hours. This flood of new manufactured chemicals caused severe dislocation in the workforce traditionally involved in bleaching fabrics, but rapid chemical bleaching also led to dramatic decreases in the price of whitened cloth.[13]

Alkali, including sodium carbonate (soda ash or washing soda), had traditionally been made from burned plant material, and vegetable ashes were sometimes traded internationally. In the 1790s, however, the Frenchman Nicolas Leblanc developed a process for making alkali in a factory using a combination of minerals and coal. His first step was to dissolve common salt, sodium chloride, in strong sulfuric acid to produce sodium sulfate, a process that also produced corrosive hydrogen chloride gas which was the most damaging by-product of his process. Next, he crushed the sodium sulfate together with calcium carbonate (lime), and coal. He then burned this complex mixture to produce usable sodium carbonate, waste calcium sulfide, unpleasant smelling hydrogen sulfide gas, smoke and soot. In 1791 Leblanc opened a factory in France and for two or three years was able to make and sell as much as 300 tons of sodium carbonate per year. In 1794, however, Leblanc fell afoul of the French Revolution, and despite having a government patent on his alkali process, his factory was seized, he was put out of business, and his trade secrets were widely publicized.[14]

In the early-nineteenth century, limited factory production of alkali continued in France, but the center of the industry soon shifted westward to Ireland and Britain, where James Muspratt became one of the leaders of the expanding chemical industry. In 1818, he and a partner had begun manufacturing materials like acetic acid, hydrochloric acid and potassium ferrocyanide (prussiate of potash) in a factory in Ireland. In 1822, however, sensing better business opportunities, Muspratt opened a factory in England near Liverpool and by the next year had begun to focus on using the Leblanc Process to make *alkali*, especially sodium carbonate. (Sodium carbonate was one of the essential ingredients in glass making, while potassium carbonate, another alkali, was needed for the manufacture of soap and fertilizer and was also useful in glass making.) Muspratt enjoyed considerable success and then went on to open several other Leblanc soda works at different locations in England. Over the next thirty years, with Muspratt's factories in the lead, the total annual production of sodium carbonate in Britain climbed continually to 140,000 tons, supporting greatly increased production of textiles, glass, soap, and so on.[15]

The Leblanc Process was the best technology available at the time, but it was still extraordinarily dirty and inefficient. It produced huge amounts of waste and pollution to the point where factories employing it often blighted their entire neighborhood, killing most vegetation, including farm and garden crops, causing respiratory distress, and even damaging buildings due to air pollution. The disposal of solid wastes from these factories also spread huge quantities of pollution farther afield. In one area, for example, about fifteen miles up-river from Liverpool, nearly a square mile of marshland was filled with waste piles of calcium sulfide (galligu) many feet high.[16]

To make eight tons of sodium carbonate, the Leblanc process had to create seven tons of solid waste and five and a half tons of hydrogen chloride.[17] Most of the solid waste consisted of calcium sulfide, but mixed in with it was unburned coal and toxic metals, including arsenic. The air pollution produced by the hydrogen chloride was so bad that even before the passage of any environmental laws, Muspratt's companies were continually being sued for creating a public nuisance, and they eventually had to take steps to reduce their toxic discharges.

In 1863, after decades of tolerating extreme air pollution from the Leblanc Process, the British Parliament finally passed the *Alkali Act*, the first modern air pollution control legislation. It required 95 percent of the polluting hydrogen chloride gas coming out of the Leblanc process to be captured. At the time, this could only be done using a kind of primitive "scrubber" that had been invented in 1839 by another alkali manufacturer, William Gossage. Gossage's method passed the gas through a tower packed with coke or charcoal, with water continually trickling through it, converting most of the hydrogen chloride into liquid hydrochloric acid. This dramatically reduced the air pollution, which was felt to be the most serious problem, but because there was little market for hydrochloric acid, factories simply disposed of it by dumping it into the nearest body of water, generally killing all the fish and other aquatic life in the process. Alkali manufacturing was the first part of the chemical industry to cause truly widespread environmental pollution, and also the first to experience meaningful regulation, although early on little was done to control either the land disposal of the calcium sulfide wastes, or the toxic discharges of liquid hydrochloric acid.[18]

Organic Chemistry

As scientific chemistry began to come into its own after 1800, it also found itself wrestling with the ancient question of whether living things were fundamentally different from everything else in the natural world. It was known that many different compounds that could be derived from animals or plants had never been detected in common rocks and minerals, and such materials were known as *organic substances*. In fact, most alchemists and early chemists believed that it was impossible to make such substances in the chemical laboratory because they could only be synthesized in a living creature possessing a *vital essence*, a belief that came to be known as *vitalism*. Organic molecules usually contain carbon and hydrogen, often combined with other atoms drawn from the other four most common elements present in living things—oxygen, nitrogen, phosphorus and sulfur. This makeup ultimately reflected the most basic chemistry of earth's plants in which the creation of carbohydrates through photosynthesis continually provides both energy to ecosystems and several of the raw materials for the synthesis of the other classes of organic molecules including proteins, fats and nucleic acids.

Following the advances made by Lavoisier, John Dalton and others in the early decades of the nineteenth century, chemists had begun to seriously question vitalism and took it as a challenge to try to synthesize organic compounds from purely inorganic starting materials. One of these was the German Friedrich Wohler. In 1824, starting only with mineral-based substances, he succeeded in synthesizing oxalic acid, a substance commonly found in many plants, including spinach. By 1828, he had also synthesized urea, a material present in mammalian urine that many believed could only be made in animals. Over the following decades, successful laboratory syntheses led to more and more doubt about the importance of a vital essence in making organic molecules, especially after 1847 when another German chemist, Hermann Kolbe, succeeded in making acetic acid—the sour substance in vinegar—from inorganic starting materials.

By the mid-1850s, enough progress had been made for chemists to begin to feel that it was well within their power to synthesize organic substances from completely inorganic starting materials. At that point, the talented French chemist Marcellin Berthelot began a sustained campaign to disprove the doctrine of vitalism once and for all. Berthelot was convinced that everything in nature, whether living or not, obeyed the same basic physical laws, and that the chemistry of organic substances was not fundamentally different from that of inorganic ones. In a series of brilliant experiments beginning in the 1850s, he demonstrated that many additional organic compounds, including methanol, ethanol and benzene, could be made from inorganic starting materials. Berthelot also showed that by varying his starting materials, he could even create *entirely new molecules that did not exist in nature*, and in this way, he helped launch the revolutionary era of modern synthetic chemistry.[19]

Berthelot's insight represented an enormous leap for both science and technology, and marked the start of an *industrial revolution* in chemistry. From the 1850s onward, and gathering pace right up to the present, pure and applied chemists have offered the world millions of new substances with an endless array of uses, creating the bases of many technologies and transforming daily human life. Billions of tons of chemical materials came to be produced, ranging from fuels and fibers to plastics, rubber, soaps, detergents, fertilizers, glues, inks, paints, pesticides, dyes, cosmetics, and on and on. Fortunes were made, and factory production allowed many new products to be produced at prices that were affordable to broader sections of society. In a few cases, the acute toxicity of individual synthetic substances led to early limitations on their manufacture and use. But in many other cases, harmful new substances were created and extensively used before it became understood that they were causing serious damage to human health and natural living systems.

Organic Fuels: Wood and Charcoal, Coal and Coke

Wherever climate offered sufficient warmth and rainfall, trees had provided fuel for hundreds of thousands of years. Most wood was burned as is, but at some point people also discovered that wood too could be transformed by fire beyond the normal processes of rapid combustion that occurred in a campfire or fireplace. The trick was to restrict the amount of air available to the flames by partially covering the burning wood with dirt or stones. When this was done, logs would smolder for days, yet still not completely burn up; instead of producing ash, the smoldering resulted in *charcoal*, a lightweight fuel that was nearly pure carbon. Considerable air pollution was also produced.

Charcoal was hard to ignite, but once lit, it burned with a much hotter, cleaner flame than wood, and its fire was particularly well-suited to extracting metals like iron from rocks. By 1,000 BC the Iron Age had begun, and from then on until recent centuries, charcoal was an important source of specialized fuel, eventually contributing to the over-cutting of forests in many parts of the world. In England, for example, shortages of wood, and thus of charcoal, began to appear by the 1600s due to runaway tree-cutting to provide wood for ship-building, construction and fuel.

In contrast to wood and charcoal which could be made wherever trees grew, mineral coal generally had to be dug out of the earth. People had utilized it as fuel for thousands of years wherever it was available near the earth's surface, and it was especially valued in places where wood was hard to obtain. By around AD 800 in China, where early population growth had caused extensive deforestation, the same techniques used to make charcoal had also begun to be used to smolder coal in order to produce *coke*, a fuel that resembled charcoal in burning hotter and more cleanly than its parent material. (Its production, of course, also generated significant local air pollution.) Knowledge of early Chinese coke production is not known to have reached Europe, but by the early 1600s, the technique seems to have been independently rediscovered in England. Once successful coke production began there, it also came to be substituted for charcoal in the refining of iron ore, which traditionally had been smelted in charcoal-fired ovens. As iron production increased tremendously in the 1700s and 1800s, coke production—and attendant air pollution—grew along with it.

Early coke manufacture focused solely on making the coke itself, and viewed the smoke, gases and tar-like liquids produced by the smoldering as waste products. Towards the middle of the seventeenth century, however, Johann Joachim Becher, a German alchemist and chemist, became interested in the thick liquids produced by smoldering coal, and by the 1660s, he developed a process for combining coke production with the extraction of liquid *coal tar*, a dark, heavy material that he thought might be a useful product. Becher had probably been inspired by the centuries-old Scandinavian practice of smoldering pine stumps to produce *pine tar*, a material that was widely used for waterproofing and preserving ropes and wood in and around ships and boats.

By the end of the 1700s, intense interest also became focused on the gases that could be produced from coal. Instead of smoldering coal by allowing it to burn with limited oxygen, methods were developed to *distill* it by heating it in sealed containers with no significant combustion occurring. In this process, liquid coal tars collected on the bottom, coke was produced in the middle, and large quantities of hydrogen, carbon monoxide and other flammable gases were produced at the top. These gases were then carefully drawn off and could be piped to lamps around towns and cities to be burned for illumination—*gaslight*. Gaslight was in fact the most popular form of community-wide lighting in the industrialized world in the nineteenth century, and *gasworks*, or factories where the gas was made, were built in hundreds of towns and cities. Gas lighting brought great social benefits, and the coke produced could also be burned as fuel. But the huge amounts of coal tars that resulted overwhelmed any need society might have had for them, and most early coke plants made little effort to safely contain or use the coal tar. Instead, they dumped it on the landscape or into water bodies where it caused a lot of damage. In some places, nineteenth century coal tar residues have even remained significant sources of pollution right into the twenty-first century.[20]

Origins of Coal and Oil

The history of the earth includes the formation of coal, oil and natural gas over long periods of time through the breakdown of trapped layers of dead plant and animal materials that are deposited underwater and then covered over with layers of sediment. Often, this happened in areas where the land was slowly sinking. Coal beds, for example, are primarily the remains of vast areas of low-lying freshwater swamps that were very widespread in the past, especially during the Carboniferous Period, 359 million to 299 million years ago. In those days, endless generations of plants rapidly grew, died and accumulated as peat under water. The swamps where this occurred never dried out because the land underlying them continued to sink due to geologic forces. In such environments, standing water dramatically reduced the amount of oxygen available to microbial decomposers, while partial breakdown of fallen vegetation produced acids which tended to accumulate in the oxygen-poor waters. These two factors worked together to strongly inhibit bacteria and other breakdown organisms with the result that dead plant material accumulated much faster than it decomposed.

Eventually, as the land continued to subside, thick beds of undecomposed peat became covered with mineral layers such as sand, silt and clay. As those overlying sediments gradually turned into rock, they locked the organic matter in place, sealing it off completely from the atmosphere. Once this happened, the weight of the overlying layers of sediment slowly compressed, heated and squeezed water out of the peat over millions of years, finally turning it into coal along with varying amounts of natural gas. Coal comes in many different grades. Generally speaking, deposits that have been buried longer and deeper are harder, cleaner-burning and with a higher carbon content, while those that have experienced shorter term, shallower burial, tend to be softer, *dirtier* and contain less carbon (see Chapter 7).

The development of petroleum follows a related path. Here, the starting materials are not land plants growing in swamps, but algae and zooplankton living in the ocean and other large bodies of water, creatures that have short lifespans and rapid rates of growth. In such communities, enormous numbers of cells are dying and being born at any given time with the dead slowly drifting down to the bottom. Once there, the organisms sometimes accumulate in the mud under conditions where low oxygen inhibits decomposition. And as in the coal swamps on the land, these layers of marine sediment gradually become thicker and thicker with greater and greater weight of additional organic and inorganic debris. Over many millions of years, the remains of the buried dead creatures are slowly transformed by heat and pressure into petroleum and natural gas. Depending on the original mix of organisms and on the varying conditions they experience following burial, different proportions of oil and gas result. And like coal, petroleum comes in many different variations with different content of sulfur and other chemical elements in addition to hydrogen and carbon.[21]

Coal Tar Dyes

The production of coal tar as a waste product from the manufacture of illuminating gas made it readily available to experimenters, and by the 1830s, a German chemist, Friedlieb Runge, was able to isolate a material from it that created a strong blue color when treated with chloride of lime, a common bleaching agent. The shade of blue Runge produced was similar to that of traditional denim jeans, which had originally been colored with natural dye extracted from indigo plants. By the 1840s, another German chemist, August Wilhelm von Hofmann (1818–1892) showed that the substance derived from indigo and the one derived from coal tar were in fact the same.

Following Hofmann's pioneering work, chemists focused even more attention on coal tar, that dark and mysterious mixture of materials which coke and gas plants were producing in such large quantities. In 1845, Hofmann accepted a permanent position as professor and director of a new institution in London, the Royal College of Chemistry, which had been established to advance education and research in applied chemistry. There he carried out a range of research including some extending his earlier studies on coal tar. In the mid-1850s, Hofmann assigned what he thought was an interesting problem—the synthesis of quinine from the coal-tar derivative aniline—to one of his abler students, a young Englishman named William Henry Perkin. At the time, quinine was an expensive natural plant extract used to treat malaria, and was much in demand as more and more Europeans found themselves living in tropical countries.

In 1856, with Professor Hofmann abroad on a trip, the eighteen-year-old Perkin, who had only been studying chemistry for three years, was experimenting with aniline in a primitive home laboratory. By some accounts, he was trying out some different approaches to the quinine synthesis from those suggested by Hofmann. During one of these trials, he combined aniline with an oxidizing agent, potassium dichromate which also contained some toluidine, to produce a black solid in what at first seemed to be an unsuccessful experiment. But when he used alcohol to try to clean up the black material, something in it dissolved in the solvent producing a beautiful purple color. Perkin was excited by this result and soon determined that silk dyed with his new chemical would neither bleed when washed nor fade quickly if exposed to light. Quickly realizing its value to the textile industry, Perkin sought a patent for it, which he received the same year. He had discovered the first effective synthetic organic dye, and he named it mauveine.[22]

Professor Hofmann soon returned from abroad, and being primarily an academic chemist, he discouraged Perkin from focusing too narrowly on his new purple dye. Perkin, however, believed strongly in his own discovery, and soon decided to leave his job at the Royal College of Chemistry in order to set up a dye manufacturing plant in collaboration with his father and brother. It was difficult for him to establish his own business; he had no experience of manufacturing, and industrial chemical production was still in its infancy. They also had trouble obtaining the large quantities of necessary raw materials, especially aniline and nitrobenzene, that they needed to make the dye.[23] Yet in spite of these difficulties, in 1857 Perkin began manufacturing mauveine (later mauve) and selling it to the textile industry as *Tyrian purple*; it proved to be a very popular color, and brought Perkin financial, professional and social success.

After mauveine, one new dye color after another was developed through chemical modification of aniline—blues, purples, reds, yellows—with August von Hofmann himself going on to discover two new reddish dyes, rosaniline and quinoline red. These innovations transformed the textile industry. Brightly dyed cloth and fibers now became available at relatively low cost, displacing the rare and expensive natural dyes which had been the only option in the past for coloring textiles, and making available to the public a pleasing selection of brightly colored garments and linens. Chemical manufacturing in general also received a great boost as factories making dyes developed new and more efficient ways of carrying out the complicated sets of reactions necessary to create the new colors. By the 1860s, synthetic chemical dye industries had sprung up across western Europe with dozens of plants in different countries, brightening daily life, stimulating national economies and spurring the development of local chemical industries.

The economic activity surrounding the synthetic dye industry was generally welcomed by local and national governments, and the reigning belief in progress through industrialization dampened most criticism of the environmental damage that the new factories were causing. We saw how alkali production in Britain, where the industry was most highly developed, had caused horrific air pollution (along with toxic dumping on land and in the water) until air pollution legislation finally began to be passed in the 1860s. On the European continent, chemical works and metal refineries also caused significant air pollution, although even greater environmental damage was likely done to streams and rivers. In inland areas of Europe, freshwater fish had been a part of local people's diets for thousands of years, but in the nineteenth century, one stream after another saw its traditional fishery wiped out or heavily damaged by water pollution. Growing population and increasing urbanization were also causing more and more human waste to be dumped untreated into rivers—often the same rivers from which drinking water was taken—and to this were now added chemical discharges, especially from the fast-growing dye industries. This was an era when neither ecology nor scientific medicine had yet developed, and society found itself completely unprepared to monitor and assess the damage that some of the new synthetic chemicals were causing. The result was that any full accounting of the costs that accompanied the benefits of the new dyes was left to later generations. The state of chemical technology also contributed to the problem. Under the crude conditions present in those decades, many toxic and cancer-causing substances were produced accidentally in the dye industry, and a number of them caused deaths and illnesses among workers, while others destroyed life in the waters into which factories generally discharged their liquid wastes.[24]

Nineteenth-century dye-making produced a whole range of toxic and harmful by-products. Coal tar—though mild solutions of it are still sometimes used externally in shampoos and skin treatments—is itself a toxic, cancer-causing substance. Aniline is toxic if its vapors are inhaled, and early methods of making it also involved the use of highly poisonous arsenic acid. The dangers of chemical discharges containing arsenic were known as early as the 1860s after several deaths occurred near dye plants, but throughout most of the following hundred years, largely unregulated water dumping still remained the major disposal mechanism for toxic chemical discharges. In some cases, harmful by-products of industry did get transported overland for dumping into the ocean. There they were diluted but still caused damage to coastal ecosystems. Most wastes, however, were dumped directly into rivers and streams close to the factories where they were produced.

In southwestern Germany, where the national dye and chemical industries were centered, the ecosystems of the Rhine and other major rivers were essentially killed off one by one. Industrialists and their allies in government even publicly acknowledged that they viewed the nation's rivers as sewers used to transport industrial wastes to the sea. Hundreds of miles of streams were devastated in this way, a pattern that was later to be repeated around the world wherever large-scale, chemical-based manufacturing took hold. Even after arsenic oxide came to be replaced by nitrobenzene in the making of dyes, arsenic pollution from chemical manufacture still persisted as a by-product of the manufacture of sulfuric acid. And as late as the twentieth century, the synthesis of aniline often produced several toxic by-products including napthylamines, which greatly increased workers' chances of developing bladder cancer.[25]

Expansion of Chemical Production

As the nineteenth century wore on, the list of manufactured chemicals grew longer and longer. By 1900, many different kinds of acids, alkalis, fertilizers, bleaches, dyes, explosives, photographic chemicals, early synthetic fibers (rayon from cellulose in 1898), synthetic perfumes, artificial flavors and food preservatives were all being produced by the chemical industry. These were soon to be joined by early synthetic plastics like Bakelite (1908), a substance which quickly found widespread use in the rapidly expanding electrical industry. Chemists were also involved in developing ways of refining aluminum from ore; in fractionating air to produce industrially useful gases like nitrogen, oxygen and carbon dioxide; in hydrogenating oils through the use of catalysts to make margarine as an inexpensive substitute for butter; in vulcanizing rubber to make it flexible and crack-resistant; and in making paper through the *pulping* of wood fibers. Despite generating massive pollution, the rush of new materials stimulated many different kinds of manufacturing and led to a great expansion in consumer products. Both were important drivers of economic growth.

Throughout the nineteenth century and into the twentieth, chemists and chemical engineers continued to produce a greater and greater variety of synthetic substances. By the end of the twentieth century, *tens of millions* of different synthetic compounds had been created in laboratories and factories, joining the large number of naturally occurring molecules that had also been identified (110 million organic and inorganic chemical substances are officially listed by chemists, the majority created for the first time in the last two centuries, with 15,000 new ones added every day).[26] By the year 2000, roughly 70,000 different chemicals were in commercial production around the world. Industrialized nations like the United States annually produced large quantities of 11,000 different substances, and more than 3,300 different chemicals were used every year in quantities greater than one million pounds.[27]

This dramatic expansion also had a human dimension. Thousands of chemists and chemical engineers—people who design and supervise chemical processes and chemical production on an industrial scale—were recruited, trained and then devoted their careers to the efficient production of a wide array of natural and synthetic chemicals. Many of the manufactured chemicals they produced, both natural and synthetic, turned out to be either safe or not particularly harmful at the exposure levels people and animals commonly experienced. But during the course of the twentieth century, the total number of chemicals in use increased to a point where toxic effects from even a small fraction of them had the potential to cause significant harm. In many cases, the damage caused was not immediately obvious, but showed up later, sometimes far away from the site where the chemicals were originally made or used. Nor were individual toxic substances the only problem. Combinations of two, three or more chemicals, which might have even originated from different sources only to combine by accident in our bodies or in the environment, can sometimes interact to produce new harmful effects beyond those of any single substance.[28] Most of these chemicals are undetectable by our senses, but this does not stop them from causing damage to human health or to the natural ecosystems on which human survival ultimately depends.

Petrochemicals

In the ninth century, Middle Eastern alchemists were the first to use careful distillation to produce kerosene, tar and other products from petroleum. They did so on a very limited scale, however, and much of their technology was eventually lost. A thousand years later, in the mid-nineteenth century, liquid fuels once again began to be manufactured from coal and oil. In the 1840s, a talented Canadian physician and geologist, Abraham Pineo Gesner, extended the existing techniques used for the extraction of coal tar to create a flammable oil from coal that he called *kerosene*. In the following decade, two Polish researchers, Jan Zeh and Ignacy Lukasiewicz, discovered a more economical process for manufacturing kerosene from petroleum, and Lukasiewicz then went on to develop the first modern oil wells, and in 1856 the first modern oil refinery. He also invented an improved kerosene lamp.[29]

In 1850, there had been no petroleum industry, but less than a century later, several of the world's largest businesses were oil companies. Oil's impact was so great that by the 1920s, the leading edge of innovation in the overall chemical industry had shifted to *petrochemicals*—substances derived from petroleum—and during the twentieth century, petroleum products came to play central roles in manufacturing, transportation, agriculture, construction, consumer products, and military operations. The first mass-produced petrochemical that was neither a fuel, lubricant nor paving material was isopropyl alcohol, a solvent and cleaning fluid that was placed on the market by Standard Oil of New Jersey in 1920.[30] (The environmental impacts associated with the extraction, refining and transport of petroleum will be described in Chapter 7.)

Oil Refining

The original starting materials for oil and coal always included a wide variety of living things whose bodies contained an enormous range of biological chemicals. This was true whether the starting material was swamp vegetation and all the animals and microorganisms associated with it, or marine plankton and the accompanying sea life. Over millions of years, the original diversity of compounds was reduced somewhat as heat and pressure transformed the starting materials into fossil fuels, but even then, the resulting coal and oil deposits remained immensely complex mixtures of molecules, often containing hundreds of different substances. For human technology, this offered great opportunities in the form of a wide range of useful materials extractable from fossil fuels, but it also presented some of industrial chemistry's greatest challenges due to the difficulty of separating out pure substances from such complex mixtures.

The first step in petroleum refining always involves slowly heating the crude oil without air to separate out different components, a process known as *fractional distillation*. When this is carefully done, different components boil off from the oil at different temperatures and can be collected. Often, however, the fractions which result are themselves mixtures of substances which cannot be used directly as finished products. Fractional distillation works because most smaller molecules are lighter in weight and boil at lower temperatures than larger, heavier molecules. The smallest molecules present in crude oil, the ones with the fewest carbon atoms, are gases—methane, ethane, propane, butane—the names are familiar ones in the world of cooking, heating and consumer products. Slightly heavier components tend to be liquids, with lighter liquid fractions going for gasoline, medium fractions being turned into products like diesel and home heating oil, and heavy liquids being processed for lubricants and for heavy fuel oils for ships and power plants. After all of these have been boiled off, a mixture of solids, semi-solids and ultra-thick liquids remains, from which refiners learned to extract products like tar, asphalt, wax and petroleum coke.[31]

To an academic chemist, all of these fractions are interesting. But to an industrial chemist or a chemical engineer, the fractions with the highest economic value are the top priorities. Throughout the twentieth century, for example, demand for gasoline grew dramatically as internal combustion engine cars turned from rare curiosities into mass-produced consumer products (see Chapter 6). The fuel they require is blended from a complex mixture of petroleum fractions, with the majority of components being compounds containing seven to ten carbon atoms, such as heptane, octane, nonane and decane. Diesel fuel, which is heavier and is commonly used in trucks, boats and farm tractors, as well in space heating and emergency power generation, is made up of larger molecules, primarily those containing ten to fifteen carbon atoms.

Chemical engineers working for industry responded to economic incentives by trying to produce as large a proportion of high-value petroleum products as possible. Fractional distillation and secondary separations allowed significant quantities of marketable substances like those used in gasoline to be extracted directly from crude petroleum, and there was also often a market for the lightest gases that could be burned as fuel. Yet a large part of most barrels of oil still consisted of components that were either too heavy for such uses or had molecules of the wrong shape to be well suited for use as gasoline or other valuable products. In response, a number of highly talented chemists and engineers little by little invented a series of processes that made it possible to transform the original mixture of substances found in crude oil into products of greater economic value.

In the most general terms, they did this in three ways: by breaking up larger molecules into smaller ones; by combining smaller molecules into larger ones, and by converting molecules of a given shape into a different configuration. Often, a final product was produced by first changing the size of the molecules, and then by altering their structure to a more desirable form. The original light, liquid crude oil fraction, known as naptha, for example, burns poorly in internal combustion engines due to its abundance of straight-chain hydrocarbon molecules. To improve its combustion, refiners learned to chemically alter the original molecules into ones of roughly the same weight but with very different molecular shapes—branching or ring-shaped molecules generally burn better than straight-chain ones.

The first set of techniques in which larger petroleum compounds are broken into smaller, more useful ones, is commonly known as *cracking*, and the first successful patent for a process to do this was obtained in 1891 by the Russian engineer, architect, and inventor Vladimir Shukhov. Other important advances were made in the 1910s and the 1930s, and there have been gradual but steady improvements ever since. Basically, refineries use different combinations of heat, pressure and steam, in conjunction with catalysts, to carry out these conversions, employing processes with names like *thermal cracking, steam reforming, fluid-bed cracking, hydro-cracking*, and so on. These provide ways to maximize the economic value in each barrel of crude petroleum by producing hundreds of different products that can be used as fuels, lubricants, chemical reagents, paving materials, personal care products and pharmaceuticals. Among the most important substances chemists learned to make from petroleum are *chemical feed stocks*, materials which form the basis for the manufacture of plastics, detergents, glues, paints, pesticides, and a wide range of other products. To achieve such high levels of technology, modern oil refineries are some of the most complex, expensive and highly automated of all industrial facilities, costing billions of dollars to build, and sprawling over large sites. Many operate continuously, 24 hours a day all year round with only an occasional shut-down for maintenance.[32]

Plastics and Plastic Pollution

The early decades of the twentieth century saw the chemical treasure chest of fossil fuels—especially petroleum—open wider and wider. Many components in crude oil were identified for the first time, and many thousands of new synthetic molecules were created. Not all of these turned out to be very useful or of great economic significance, but some proved to be very important for technology, while a fraction also ended up causing serious damage to human health and/or the natural environment. In many cases, new materials derived from petrochemicals came to be used in place of substances derived from natural products. *Plastics*, for example, are materials which can assume a wide variety of shapes, and synthetic plastics are usually made from organic starting materials— those containing hydrocarbon or carbohydrate molecules—often with the addition of other chemical elements such as nitrogen, chlorine or silicon. Plastics are polymers, big molecules assembled by linking together large numbers of one or more types of smaller molecules. Most can be shaped into a wide range of useful objects by either being molded or squeezed out through dies (extruded). Petrochemicals are the source of synthetic plastics, but plastic-like materials have also long-existed in nature, with most land vertebrates producing a natural plastic material based on the protein keratin. In mammals it is used to form a wide variety of different structures—hair, fingernails, horses' hooves and rhinoceros' horns are a few examples. In birds, it is used to make feathers and bill covers. In each case, it is laid down by body tissues in such a way as to produce structures of widely varying textures, sizes and shapes.

The first commercial plastic was celluloid, a material developed in the nineteenth century from chemically treated plant cellulose. After its discovery in the 1850s, it became a commercial success in the 1870s, and for several decades was commonly used for molded domestic items like combs, brushes, makeup boxes and picture frames. By the end of the nineteenth century, chemically rearranged cellulosic fibers that resembled silk also began to be marketed as *rayon*, and in the 1920s, another polymer, cellulose acetate, also began to be manufactured for use as a substitute for silk. These materials still have limited uses today, but by the 1930s, industrial chemists had figured out how to make several entirely new kinds of synthetic fibers—nylons, polyesters and acrylics—and by the 1940s, these were all being made into textiles, where they have largely supplanted manufactured cellulose-based fibers. Natural fibers like cotton and wool still remain popular, but since the 1930s, petrochemical-based synthetics, especially polyester, have played ever more important roles in the world of textiles. And fibers were not the only uses to which these entirely new materials were put. Acrylics, for example, form the finished surface in many modern, water-based paints, and are also used to make solid clear plastics.

Plastics are everywhere nowadays. Products as different as household sewer pipes and moderately priced clarinets, for example, are commonly made from a plastic called acrylonitrile butadiene styrene, or ABS. (Over-harvesting and inadequate forest management has made the tropical hardwoods like rosewood that were traditionally used in clarinets and other instruments so scarce and expensive that it is now used only for elite musical instruments.) Dozens of other plastics are also used in a wide range of specialized consumer and industrial products—for example, poly (poly methacrylate), polyamide-imide, polyether etherketone, polyphenyl-sulfone, polybutylene terephthalate and polyoxymethylene, although these materials are not familiar to the general public.[33]

In contrast, a half dozen kinds of plastic are so much a part of everyday consumer products that they have been given familiar recycling numbers to aid in their separation in the solid waste stream:

1   Polyethylene terephthalate (PETE or PET), which is used to make water bottles, food packages and polyester clothing fiber.
2   High-density polyethylene (HDPE), from which stronger bottles are made like those used to pack liquid laundry detergent and milk.
3   Polyvinyl chloride (PVC), which, in addition to plumbing pipes, is used in house siding, lawn furniture, clear product packaging, shower curtains and flooring materials.
4   Low-density polyethylene (LDPE), which is frequently used for plastic bags and food storage containers.
5   Polypropylene (PP), which is also widely used in food packaging.
6   Polystyrene (PS), which is used to produce cups, food trays, cushioning packing material, thermal insulating sheets and solid plastic tableware.

Polystyrene is often *foamed* (expanded with gas) to produce rigid foamed plastics as for disposable coffee cups.[34] Other important and widely used plastics include polyamides, a chemical grouping containing important synthetic molecules like nylon as well as some natural materials like silk and wool; polyurethane, which is used to make varnishes, synthetic foam rubber cushions and cushioning gel pads; and polycarbonate, a hard, clear, shatter-resistant plastic used for the lenses of safety glasses and automobile headlight covers (see "Bisphenol A" in Chapter 5).

Familiar plastics like these, if chemically pure, pose little direct health risk. Most are quite stable, at least in the short to medium term, and don't tend to interact strongly with living tissue. But most commercial plastics are not chemically pure. Instead, they contain fillers, additives, plasticizers, un-reacted starting materials and impurities, many of which *are* toxic or hazardous in varying degrees, a topic to which we will return shortly. Synthetic plastics are also entirely new to the earth's environment, which means that natural decomposers, including bacteria, have never evolved enzymes to digest them. In the short term, this quality offers some advantages because in addition to relatively low cost and moldability, many plastics are durable, resistant to decay and capable of being attractively colored and textured. In all but the driest climates, for example, unpainted wood fencing or house siding will rapidly begin to show signs of decay due to attack by fungi and bacteria, while commonly used polyvinyl chloride siding, if protected by ultraviolet resistant additives such as titanium dioxide, can remain serviceable for decades. (Ultraviolet light from the sun generally tends to physically degrade both wood and pure plastic alike.)

The flip side of resistance to decay in synthetic plastics, however, is potential long-term pollution from broken or discarded items, which in some cases may take thousands of years or longer to fully decompose. The most extreme example of this problem can be seen in the middle of the northern Pacific Ocean, thousands of miles from land, where ocean currents bring together plastic debris from a vast area. Known as the Pacific Trash Vortex or Great Pacific Garbage Patch, this area consists of billions of billions of tiny plastic fragments covering perhaps as large an area as the United States. In the ocean, ultraviolet light and physical impacts break up larger pieces of plastic into nearly microscopic bits which float indefinitely near the surface. The patch has been developing for decades and now contains as much as 100 million tons of debris, four fifths of which is thought to have been discarded along coast lines in Asia and North America, with the rest coming from ships at sea. Over time, the plastic particles tend to become smaller until they are so tiny that they begin to be ingested by small marine creatures. In this way, they enter the food chain where they are passed along from species to species, eventually reaching fish which are then caught for human consumption.[35] The long-term impacts of this debris on people and the rest of nature will not be clear for some time. In 2010, the results of a twenty-year research program also documented the formation of a second large area of floating plastic trash in the Atlantic.[36] And on land, the amount of plastic in landfills and littering the environment dwarfs even the mountainous quantities of it in the great ocean garbage patches; it is estimated that since the 1950s, one billion tons—2,000,000,000,000 pounds—of plastic have been discarded, and that only a small fraction of that has been recycled.

Chlorinated Hydrocarbons

Once the synthesis of new molecules became common in the middle of the nineteenth century, chemists tried out thousands and thousands of different reactions, sometimes to test particular hypotheses, sometimes to simply try out new chemical combinations. In this way, they were able to create an endless array of new substances along with synthetic versions of many useful natural chemicals. Many successful experiments involved elements that came from either the left or right hand sides of the *periodic table of the elements* because it had turned out that a number of very reactive atoms (such as hydrogen, sodium, potassium, calcium, oxygen, nitrogen, chlorine, bromine and fluorine) could be found in those locations. On the right hand side of the Table, chlorine in particular proved particularly useful, as it was highly reactive and could readily be obtained from common salt—sodium chloride.[37]

Starting in the nineteenth century, large numbers of new chemicals were synthesized by combining chlorine with a wide range of other molecules, and many of the products turned out to be usable as bleaches, plastics, herbicides, insecticides, fungicides, electrical insulators or industrial solvents. As a broad class, one large array of new molecules that could be produced in this way were the chlorinated hydrocarbons or organochlorines. Some of these substances turned out to be relatively safe to use—the artificial sweetener sucralose (currently marketed as *Splenda* and under several other commercial names) has been approved for human consumption in most industrialized countries, and has not produced any notable harmful effects in more than a decade of use. But a disturbing minority of organochlorines have proved to be either extremely toxic to people, very damaging to natural ecosystems, or both. Here, we will consider a few examples starting with the commonly used substance polyvinyl chloride, the world's third most widely produced plastic. Tens of millions of tons of it are manufactured every year around the world, and it presents potential hazards at many stages of its life cycle.

Polyvinyl Chloride

To make sense of the structure of polyvinyl chloride, it is helpful to look once more at some of the key steps in the development of organic chemistry, particularly in the nineteenth century. In that period, pioneering organic chemists worked out the structure of a wide range of substances that could be derived from fossil fuels. One important group of these were simple, straight chain hydrocarbons, containing a single line of carbon atoms with hydrogen atoms bonded to them. Methane was the simplest of these, a naturally occurring gas often produced by microbes living without oxygen in sediments underwater, and also present in oil and gas deposits. Each methane molecule has a single carbon sharing just one pair of electrons with each of four hydrogen atoms. The next heavier substance in methane's chemical group is called ethane and has two carbons bonded to each other through the sharing of a single pair of electrons along with six hydrogens, three bonded to each carbon.

Under certain conditions, two of ethane's hydrogens can be lost, forming a compound called ethylene (ethene), and when this happens, the two carbons share two pairs of electrons instead of just one, and the overall molecule contains just four hydrogen atoms. Ethylene is one of the most widely produced synthetic chemicals, and is both the raw material for making polyethylene, the world's leading plastic, and also a starting material for the manufacture of many other substances. Tens of millions of tons of it are made annually around the world, mostly by the steam cracking of petroleum.[38] Once made, ethylene can undergo many different chemical reactions, one of which is the substitution of a different chemical group for one or more of its four hydrogen atoms. If just one of ethylene's four hydrogens is replaced by a different atom or small molecule, the resulting molecule is referred to as a *vinyl* group, a name chosen because it was originally produced in the laboratory from ethyl alcohol, which itself was often derived from wine; wine in Latin is *vinum*.

The first step in making polyvinyl chloride is reacting ethylene with chlorine gas to make ethylene dichloride (1,2-dichloroethane), a substance not known to occur in nature. Ethylene dichloride is flammable, toxic, corrosive and probably carcinogenic. Chemical plants go to great lengths to keep it safely contained within reaction vessels, although chemical workers still experience some risk due to its toxic nature. In the second step, ethylene dichloride is exposed to heat and high pressure which causes it to decompose into hydrochloric acid and vinyl chloride. Vinyl chloride is another completely synthetic chemical not known in nature. Heavy human exposure to it damages the liver and the nervous system and, if chronic, can cause liver cancer. The third step in PVC manufacture involves linking together large numbers of vinyl chloride monomers to make the final synthetic substance, the polymer or plastic, polyvinyl chloride.[39]

Polyvinyl chloride was first discovered in the nineteenth century, although it found no immediate practical use. As a pure chemical, it is rigid, brittle, unstable in heat and light and with a tendency to slowly degrade over time. In the 1920s, however, researchers trying to make improved synthetic rubber, stumbled upon it again and managed to find additives that could make it more durable and economically useful. Broadly speaking, these additives fell into four main categories: *plasticizers* designed to make the material more flexible and capable of being molded; *stabilizers* designed to make the material more durable and better able to withstand high and low temperatures, mechanical stress, sunlight and the presence of electricity; *lubricants* to make the plastic easier to process; and *pigments* to color it. A very tiny amount of the inherently toxic vinyl chloride monomer always remains in the manufactured polyvinyl chloride plastic, and over the years, it often finds a way to escape into the environment. Beyond this, most of the environmental impacts of PVC plastics result from the stabilizers and plasticizers that are added to them.

In the decades following PVC's rediscovery, many different additives were tried, and it was soon found that most of the stabilizers that best improved PVC's working properties at moderate cost contained lead. Lead's toxicity was strongly suspected even in ancient times, but those concerns were not enough to stop the widespread adoption and use of lead compounds as stabilizers for PVC throughout most of the twentieth century, even after many countries had begun to restrict the use of lead in paints, gasoline and most other consumer products. Eventually, however, it was demonstrated that small amounts of lead from PVC leached into drinking water distributed in plastic pipes, and also that incineration of lead-stabilized vinyl products led to discharges of lead into the air (see Chapter 5). In the European Union, these concerns led to a phase out of all lead stabilizers in PVC drinking water pipes by 2005 and in all other PVC by 2015, although there is still a large amount of toxic lead embedded in the millions of tons of PVC items in place around the world[40] (see "Lead" in Chapter 5).

Bleaching Paper and Insulating Transformers

The history of industrial paper-making also illustrates how even seemingly safe and controlled use of powerful chemicals can sometimes produce serious unintended consequences. Paper, as we know it, first began to be made in China about 2,000 years ago, and it was soon realized that light colored paper offered more contrast for writing done in black ink. This, in turn, led to a strong preference for paper made from light colored plant fibers—cotton, linen, and the like—although the relative scarcity and high cost of such fibers meant that for many centuries paper remained an expensive material. By the 1840s this all began to change after successful experiments were done on utilizing wood fibers—a much more abundant and potentially cheaper raw material—for making paper. In many cases, however, raw wood pulp, essentially shredded wood that had been chemically treated, tended to be darker than the kind of plant fibers previously used — more the color of a cardboard box than of writing paper. By the 1840s then, wood pulp for paper-making started to be chemically bleached to lighten it. In the mid-nineteenth century, this could be done using a chemical agent like calcium hypochlorite, which had been invented several decades earlier. In 1892, however, the chemical industry came up with a new type of chlorinated bleaching agent, sodium hypochlorite, a very familiar substance today because it is the active ingredient in common laundry bleach. Before long, strong solutions of something like super-concentrated laundry bleach were being widely used in paper mills. Even sodium hypochlorite, however, was not completely effective at lightening commercial wood fibers, so in the 1930s, paper mills began to combine multiple doses of hypochlorite with a single treatment of even more reactive chlorine gas.

To understand the impact of this, it is necessary to consider the complicated chemistry of paper pulp, which consists of a mixture of cellulose fibers with other substances found in trees. This great chemical complexity means that a wide range of different chemical molecules are produced in differing amounts as pulp is processed. Chemical engineers and technologists have labored for decades to maximize the output of desirable products while minimizing the production of unwanted ones, yet despite their best efforts, impurities and waste products almost always remain. The fate of the chlorine that is sometimes added to bleach the pulp offers a good illustration of this. Most of the darker colored materials in softwood pulp consists of lignin, a high polymer (plastic-like) material containing a variety of organic ring compounds that give intact wood much of its strength and stiffness. With good engineering and careful process monitoring, most of the chlorine used to bleach paper pulp ends up as harmless wastes such as chloride ion dissolved in water. But a small percentage of the chlorine generally reacts in unpredictable ways with the wide range of lignin molecules present to produce new substances that contain chlorine atoms bonded to medium-sized ring molecules. The most problematic of these accidental by-products in twentieth-century paper-making was a class of chemicals known as polychlorinated dibenzodioxins—a term sometimes shortened to *dioxins*. This is a family of compounds that can pose serious health risks to people and to other living things.[41]

Altogether, there are roughly seventy-five different dioxin compounds of widely varying toxicity. Tiny amounts of some of them are probably released naturally in forest fires and volcanic eruptions, but not at levels high enough to cause health or environmental concerns. Unintentional industrial releases, either directly from factories or as residues in products, have generally been much larger. And as with numerous other chlorinated organic compounds, the pollutants created are chemically stable and bio-accumulative, that is they tend to build up in animal and human body fat and cannot easily be excreted. (Since these substances were not normally present in significant quantities in the environment, evolution had no reason to build in ways for us to excrete or detoxify them.) Once present in our tissues—and medical research shows that all human beings now have traces of dioxins present in their bodies—they seem to have a range of damaging effects caused by their disruption of the normal ways that our cells' genes are activated or inactivated, and also of the ways that one gene influences the activity of others (see Chapter 5).

Documentation of dioxin's toxicity dates back at least to the 1970s, and by the 1980s dioxins were known to be present in paper mills, in the waste water they discharged, and in the fish living in the streams receiving the discharges. Through the natural process of bio-concentration (see "Mercury" in Chapter 5), even low levels of dioxins in the algae growing in the streams tended to accumulate at higher concentrations in the small animals that fed on them, and at yet higher concentrations in the fish which lived at the top of the aquatic food chain. And unlike many other pollutants which break down relatively rapidly in the bodies of living things, dioxins tended to remain intact for extended periods causing damage to cells. A US National Institute of Medicine's literature review found that even after seven and a half years, half of the dioxin originally present in an exposed person was still there.[42] Over the decades, the people experiencing the highest dioxin exposure were factory workers in paper mills and chemical plants—those most directly exposed to the industrial processes producing the toxic substances. But once released into the environment, the dispersal of the pollutants, followed by their bio-concentration and localization in fats (they are not water soluble, but dissolve readily in oily and fatty materials), introduced dioxin into the human food supply and even the human life cycle. Depending on the length of exposure and the concentration, people may experience short-term or long-term effects. In the short term, skin lesions and changes in the liver may result. Longer exposure can disrupt the immune system, the body's hormonal system and the development of the nervous system in infants and children. Animal experiments have also shown that long-term exposure can lead to several types of cancer.[43]

In response to growing evidence of the dangers of dioxins, the United States Environmental Protection Agency (USEPA) started limiting releases of them into the air from factories and incinerators and into water from paper mills.[44] Similar regulations were enacted in other industrialized nations, and efforts are still underway in Europe and the United States to further reduce releases of dioxin. Pressure has also been placed on the paper industry in particular to totally eliminate chlorine from their bleaching processes through the use of alternative chemicals such as the reactive form of oxygen known as ozone. Paper-making, however, is not the only activity that can release dioxins into the environment, so even forceful efforts to reduce industrial releases may lower our exposure yet not eliminate it.

According to the USEPA's data from the year 2000, the ten leading sources of releases of dioxin and dioxin like compounds in the US are: outdoor burning of trash by private individuals; incineration of medical wastes (especially items made of vinyl plastic); municipal waste incinerators; sludge from municipal wastewater treatment; coal-fired power generation; cement kilns burning hazardous waste; heavy-duty diesel trucks; magnesium smelting; industrial wood burning; and aluminum smelting. A wide range of other sources accounts for an additional 10 percent of the total. Most of the dioxins and related compounds released by burning trash are produced from PVC plastics which are used in hundreds of different products and are quite difficult to recycle.[45]

Dioxins are still present in our environment and new releases are still occurring, especially in less developed nations and in places where household trash is commonly disposed of by burning. Overall, though, society's reaction to dioxin's toxicity has largely been a success story. The EPA report describing leading sources of dioxins in 2000 also showed that in response to regulation and to modifications of industrial processes, total releases in 2000 were less than half what they were in 1995 and only a small fraction of what they had been in 1987. This offers little comfort to an individual whose health may have been damaged by these substances, but it definitely represents environmental progress. Monitoring and regulation have understandably focused on protection of human health from toxic dioxins, but in the long run, the damage done to natural systems and aquatic life may be even more important than that done to a single human generation. At least the dangers of dioxins have now been recognized, and reasonable controls on their use and release put in place in some countries.[46]

As environmental science research continued during the twentieth century, a number of other groups of synthetic chlorine-containing organic compounds were also found to be toxic and damaging to life. One such category consists of polychlorinated biphenyls (PCBs), a series of roughly 200 compounds first made in the nineteenth century but not used industrially until the 1920s, when they came into widespread service in the electrical industry. As we will see in Chapter 6, many of the most fundamental principles of electric power had been described in the first half of the nineteenth century by researchers like Michael Faraday. It had then taken another half century before many practical technologies based on those principles were developed—electrical generators, motors, lights, switches, and a wide range of devices for transporting and managing electricity in power networks. These devices included distribution lines, switching systems, capacitors, and electrical transformers which were designed to increase and decrease the voltage (electrical pressure) of current being sent along the wires. By the 1920s in industrialized countries, the electric power industry was growing very rapidly and so was busily installing electrical transmission lines along with the other elements of power grids. In order to minimize the loss of electricity being transmitted over long distances, it became necessary to step up the current to extremely high voltages after it was generated, and then, after transmission, to lower the voltage before it was delivered to customers. To do this, large numbers of transformers were required, and these soon became universal components of electrical power networks.

Early transformers were insulated electrically with mineral oil, which also served as a coolant liquid. But mineral oil was flammable, and transformers containing it sometimes caught fire. Polychlorinated biphenyls, however, were not flammable, and once they became commercially available in 1929, leading American manufacturers of electrical equipment like General Electric and Westinghouse began to use large quantities of them to insulate the inside of transformers and other electrical devices. In America, most PCBs were manufactured by Monsanto Chemical and marketed under the trade name Aroclor. They were made and sold as mixtures of related, chlorinated chemicals, and also used for a number of other purposes in addition to electrical insulation, especially in paints, inks, adhesives, flame retardants and plasticizers.

The production process for PCBs always produced mixtures of related substances. In addition to the intended compounds, manufactured PCBs also often contained small amounts of a different, related class of toxic organochlorines, polychlorinated dibenzofurans, adding yet another set of problem chemicals to the mix. In the United States, production of PCBs peaked in 1970 at 85 million pounds before its use began to be regulated and restricted, and by 1979, it had been banned in the US except for a few very specialized applications. In its place, silicone oils and other chlorine-free insulating materials were developed for use in transformers and capacitors. Their technical performance was not quite as good as PCBs, but they were free of the worst health and environmental problems.[47]

Even during PCBs' first decade of industrial use in the 1930s, chemical plant workers began to develop medical problems related to their exposure, most visibly severe outbreaks of skin lesions known as *chloracne*. In 1937, a scientific paper was published drawing connections between industrial PCB exposure and liver disease.

It was written by Cecil K. Drinker, MD, DSc, Professor of Physiology and Dean of the Harvard School of Public Health, with a number of co-authors, and was entitled "The Problem of Possible Systemic Effects From Certain Chlorinated Hydrocarbons."[48] Also in the 1930s, Monsanto Chemical licensed PCB production to foreign chemical companies which soon led to its production in several European countries and Japan. In addition, there were unlicensed factories producing PCBs in communist-bloc countries (PCBs were still being manufactured in Russia until at least 1993). Articles in scholarly journals place the total quantity of all PCBs produced between the late 1920s, when industrial production began, and the 1990s when it ended, as exceeding one and a quarter million tons, or at least two and a half billion pounds. And as late as 1981, it was estimated that 131,200 transformers containing PCBs were still in use just in the United States.[49]

As a class, PCBs were responsible for a number of major chemical pollution episodes. Between 1947 and 1977, for example, General Electric Company's two transformer manufacturing plants on the upper Hudson River in New York State dumped more than one million pounds—at least 500,000 kilograms—of waste polychlorinated biphenyls into the Hudson River, and during that same time period, their factories in Pittsfield, Massachusetts dumped a similar amount into both the Housatonic River and land areas around that town. Both locations have since been scenes of massive, multi-year clean-up efforts costing many hundreds of millions of dollars.[50]

Herbicides

Another very important group of organic chemicals containing chlorine are those used as herbicides, especially those designed to kill weeds growing in fields of grain like wheat, rice and corn. These important food plants are members of the grass family, which, in turn, belongs to a larger group of flowering plants known as monocotyledons (monocots). Monocots have that name because their seedlings start off life with one rather than two seed leaves (cotyledons). They are very important in nature, and include grasses, palms, bamboos, orchids, lilies, daffodils and the like. Planet-wide, however they coexist with the even more species-rich dicoyledons (dicots) whose seedlings start off life with two seed leaves. The dicots include most of the remaining flowering plants, from apples to zinnias, and dicot species are often the most troublesome weeds of grain crops.

During World War II, a team of British plant biologists that was probably studying possible ways to destroy enemy food crops, focused their research on *auxins*, a group of chemicals in flowering plants that control growth; they had first been discovered in the late 1920s and can be compared to hormones in animals. As part of their research, they discovered a chlorinated compound which partially mimicked the action of certain auxins. Natural auxins are produced only in small quantities in plants. But this new substance—2,4-dichlorophenoxyacetic acid (2,4-D for short)—was designed to be applied at much higher concentrations than the natural substances, and when this was done, it caused cells to grow so rapidly that the treated plants died. And for biochemical reasons that are still only being worked out, this toxic effect applied only to dicots and not to monocots, which meant that 2,4-D could be used to kill broad-leaved weeds in fields of wheat, corn, rice, barley or millet without killing the crop plants. During World War II, 2,4-D was never actually used against enemy crops, but the idea of using herbicides in war nonetheless seems to have become part of military thinking.

After World War II, 2,4-D was made available for commercial use as the first completely synthetic *herbicide*. It was inexpensive to produce and relatively non-toxic to people in the short term. Soon, it became the most widely used *weed killer* in the world, and in fact, it is still widely used today although more often in combination with other synthetic herbicides rather than all by itself (in the United States, in 2007, herbicides of one type or another were used to treat 226,000,000 acres of farmland[51]). After more than half a century of use all over the world, a large number of observations have accumulated about 2,4-D's effects on people and the environment. In general, it has proven to be relatively non-toxic for farmers and chemical workers, although some studies have linked long-term exposure to it to increases in the incidence of amyotrophic lateral sclerosis (ALS or Lou Gehrig's disease), as well as to a large family of uncommon cancers known as *non-Hodgkin's lymphomas*. Strong 2,4-D solutions can cause skin irritation and eye damage, so agricultural workers do need to be cautious when handling it. In general, though, it has low toxicity for wildlife and tends to be broken down fairly rapidly by micro-organisms in soil and water.[52] On the plus side, 2,4-D has helped to dramatically increase global grain production, lowering the chances of human starvation and hunger. On the minus side, it has also helped to encourage the explosion of mono-culture turf-grass lawns and golf courses by making it possible to easily eliminate broad-leaved plants, and this, in turn, has led to heavier use of synthetic lawn fertilizer and insecticides, many of which have ended up causing environmental damage.

In retrospect, the developers of 2,4-D were fortunate to have discovered a relatively non-toxic, low-cost substance that proved to be so useful to farmers and such a boon to the human food supply. But when chemists attempted to build on this breakthrough, their work soon led to other chlorinated herbicides that in the long run proved much less benign. After success with 2,4-D, researchers began synthesizing and testing compounds with related chemical structures. (It is often the case—although not always—that chemical substances with similar structures have similar properties.) Following this approach, one of the synthetic substances they developed was 2,4,5-trichlorophenoxyacetic acid (2,4,5-T), a molecule which has an identical carbon backbone to 2,4-D, but which features three chlorine atoms substituted for hydrogens around the carbon ring instead of two. When they tested it, they found that it too mimicked plant growth hormones, stimulated overgrowth and caused plants to die when applied at moderate dose rates. By the 1950s, it too had become a widely used weed killer.

Herbicide Use in the Viet Nam War

In the early 1960s, as an extension of the Cold War with the Soviet Union, the United States went to war in Viet Nam, where its military was soon fighting in dense rainforest terrain against a highly dispersed and often well-camouflaged guerilla army. Under these conditions, some military planners harkened back to the notion that had secretly been worked on in the dark days of World War II—to use herbicides as agents of war. This had actually already been tried with some success by the British in the 1950s in Malaya against both jungle cover and agricultural fields when they battled a guerilla force.[53] In Viet Nam, however, the original use proposed was not to destroy food crops, but rather to clear thick vegetation around military bases so as to limit the possibilities of surprise attacks. This was then tried with some success, and planners moved on to the idea of landscape scale chemical defoliation as a way of improving overall military operations, and eventually of crop destruction as a way of weakening the enemy.

The campaign began in 1962 with the herbicide spraying of areas adjacent to American and South Vietnamese military bases; later it was expanded to include the spraying of large tracts of forest throughout Viet Nam. Eventually, food crops in North Viet Nam were also targeted. Between 1962 and 1971, more than nineteen million gallons of various herbicides were released over Southeast Asia, and by far the most common formulation used was Agent Orange (named for the color of the stripes on the chemical barrels), a mixture of 2,4-D and 2,4,5-T. Over the course of nearly a decade, this vast aerial spraying exposed millions of Vietnamese and hundreds of thousands of US and allied service personnel to this herbicide mixture (US allies in the conflict included South Korea, Thailand, Australia, New Zealand and the Phillipines). Smaller amounts of arsenic-containing herbicides, known as Agent Blue, were also sprayed by the military in Viet Nam.

As we have seen, 2,4-D, by itself, is relatively non-toxic, while completely pure 2,4,5-T is at worst only moderately toxic. In theory then, the mixture of the two should have caused defoliation without directly harming people, although herbiciding of more than ten percent of the entire land surface of Viet Nam would in any case cause enormous damage to ecosystems and to plant and animal populations in many parts of the country. It turned out, however, that the manufacturing process for 2,4,5-T was much trickier than was generally understood at the time. One result of this was that under normal factory conditions, the manufacture of 2,4,5-T *always produced* some highly toxic dioxin, TCDD (2,3,7,8-tetrachlorodibenzo-p-dioxin), one of the most harmful of all the chlorinated organic compounds, and one which remained present in the final product. So when millions of gallons of Agent Orange were sprayed over Southeast Asia during the Viet Nam conflict, significant quantities of dioxin were also widely dispersed, even though dioxin generally represented less than one part in one million of the overall herbicide mixture. In the United States today, levels of dioxin in the environment which exceed one part per billion are considered hazardous and require clean-up.[54]

After the Viet Nam war ended in 1975, neither the United States nor the Vietnamese governments was willing to acknowledge the toxic effects of this unprecedented spraying. Complete understanding of the profound impacts of even low-level exposure to the mixture of chemicals employed took some time to emerge, and both countries at first sought to reassure their veterans that they were not at risk for serious health effects. As time passed, however, both countries—especially Viet Nam—came to document extremely widespread and serious health problems in the aftermath of the Agent Orange spraying.

Overall, no single, fully objective study of the total effects of the decade-long Southeast Asian herbicide campaign has ever been carried out. But enough hard facts have emerged from authoritative sources to make it possible to paint a fairly detailed picture. In Viet Nam itself, where several million people are believed to have experienced significant chemical exposure, at least hundreds of thousands have suffered very serious impacts including cancers, birth defects (often horribly disfiguring), liver disease and neurological problems such as Parkinson's disease. Long-lasting effects of exposure to dioxin and to other chemicals released by military operations are one of Viet Nam's greatest public health problems. There also remain a number of locations with dangerous, concentrated contamination, particularly at the airfields where the defoliant spray planes were based.

After the American and South Vietnamese forces were defeated in 1975, the US Government initially told military personnel and civilian Defense Department employees not to worry about their exposure to defoliants. Starting in the early 1990s, however, the US began to acknowledge that Agent Orange exposure was one possible cause of a range of diseases. Since that time, it has come to recognize all of the following conditions as potentially being caused by Agent Orange exposure: ALS, peripheral neuropathy, amyloidosis, chloracne, chronic lymphocytic leukemia, Type II diabetes, Hodgkin's disease, ischemic heart disease, multiple myeloma, non-Hodgkin's lymphoma, Parkinson's disease, porphyria cutanea tarda, prostate cancer, respiratory cancers and soft tissue sarcomas. In addition, children born to Viet Nam veterans, particularly female veterans, suffered increased risk of birth defects, especially serious neural tube problems such as spina bifida.[55] Many service personnel exposed to Agent Orange did not suffer any of these serious conditions, but thousands of others did, and nearly five decades after their exposure in the 1960s, some veterans are still struggling with medical problems traceable back to dioxin exposure.

Newer Herbicides

2,4-D has been an important component of many agricultural systems since it first came to be used after World War II. But chemical herbicides and insecticides cannot alter the basic rules of natural selection. Just as widespread use of antibiotics has unintentionally *bred* bacteria that are immune to those drugs, so broad-scale insecticide and herbicide use has, in turn, bred weeds and insects that are resistant to previously deadly chemicals. This pattern has been seen for decades and is likely to continue indefinitely. Technology's response has been to continue to develop new pesticides, and in 1958 a member of a new class of chlorinated hydrocarbon herbicides was introduced under the commercial name atrazine. This substance has now become the most widely used herbicide in the United States and much of the rest of the world, with roughly 75 million pounds being used in the US each year, primarily on corn crops (maize).[56] Atrazine is inexpensive and effective, but over several decades there have also been strong suggestions that it too can cause health and environmental problems, mainly due to its slow rate of breakdown in the environment, especially in groundwater. A number of scientific studies suggest that it imitates female hormones in animals, and that in some places this has led to reproductive failure in frog populations by sterilizing the majority of male frogs while turning some of them into hermaphrodites (see "Endocrine Disruptors" in Chapter 5). In humans, other research findings have suggested that it may cause birth defects and low birth weights, as well as menstrual difficulties for women. In 2004, it was banned in the European Union due to suspected health threats, although at the present time it continues to be widely used in the United States and many other countries.[57]

Insecticides

Ever since early farmers began to grow un-naturally large concentrations of individual crop plants, the insects that were adapted to feeding on those plants have tended to increase in number to the point where they could become a problem. Unlike farm fields, which typically are planted with just a single crop species, most natural ecosystems contain a wide variety of plant species living mixed together, and this was almost certainly the pattern when cultivation of food plants began. Land plants and insects have coexisted on earth for more than 400 million years. In the earliest days, insects may have fed primarily on dead and rotting vegetation, but for at least 300 million years, they have also fed actively on living plants, eating leaves, sucking sap, boring through stems, and attacking flowers, fruits and seeds. This, in turn, led to extensive mutual evolution, a kind of natural *arms race* which saw plants change over time so as to reduce insects' success in feeding, while insects changed to improve their physical and chemical abilities to feed on and digest plants. These intricate and numerous adjustments all happened in environments where most plant species were dispersed among large numbers of other species rather than growing all alone in pure stands as they do in a farm field. Most of the individual plant species growing in diverse, natural communities had their own specialized insects feeding on them, and the overall living community also supported a rich set of predators, pathogens and parasites—carnivorous insects, spiders, lizards, birds, fungi, bacteria, viruses—all of which helped to keep the numbers of plant-eating insects in check.[58]

As early agriculture developed, however, farmers soon learned that various individual species of grain and other crops required different methods and timing of sowing, cultivation, harvesting and processing. This, in turn, led to growing larger and larger patches of a single food plant species, all of which could be planted, cultivated, harvested and processed in the same way at the same time. From the point of view of the insects which had fed on a particular species of crop plant for thousands or even millions of years—often evolving their whole life history to match that of their food plant—this represented an unparalleled opportunity. Suddenly, large numbers of their preferred food plants were densely concentrated together instead of being widely interspersed among a large number of other plant species. This meant that there was a lot more for them to eat, which allowed the insect populations to dramatically increase, and that, in turn, may even have allowed them to evolve more rapidly. Most insect species, of course, are also impacted by fluctuations in the weather, as well as by changes in the populations of their predators, parasites and pathogens. In some years, these influences kept aphids, grasshoppers and other important plant-eaters' populations from rising to pest proportions despite the presence of larger and larger patches planted with individual crop species. But in years when these other controls failed, human food crops could be significantly affected, often raising the specter of crop failure, food shortage or outright starvation.

Early Insecticides: Sulfur and Pyrethrum

The oldest technique for dealing with too many plant-eating insects was to manually pick them off—in the case of grasshoppers, a virtue was often made of this necessity by adding them to the human food supply. With smaller insects, like aphids, however, this was not practical. One early response, within the limits of the human and animal labor available, was to expand the area of land cultivated. This, of course, destroyed even more natural habitat, and also risked generating even larger insect outbreaks. In such a situation, any techniques that might reduce the number of plant-eating insects were eagerly sought, and even thousands of years ago people tried out various substances to see if they could reduce the insect pests threatening crop plants. The first modest success came through application of compounds containing the element sulfur, which is recorded as having been used for this purpose nearly 5,000 years ago in ancient Mesopotamia.[59] In some parts of the world, pure or chemically combined mineral sulfur was available in significant quantities; pure sulfur could be burned to kill pests, while a variety of sulfur minerals could also be ground up and dusted over insects. Even today, sulfur insecticides still see some use, particularly on organic farms. They are reasonably effective against plant mites, scale insects and thrips—important pests of crops like citrus—as well as against many fungi which attack crops. They are much less effective, however, against other common insect pests like moth and beetle larvae or grasshoppers.

Land plants have existed on earth for hundreds of millions of years, and one of the key themes in their development over vast stretches of time was the evolution of a wide range of defensive chemicals which served to discourage feeding by insects. During many centuries of agricultural civilization, as farmers struggled to meet the challenge of growing enough food for a slowly increasing human population, people eventually came to identify a few plant species that were especially rich in such defensive compounds. Around 2,000 years ago in Asia, for example, people discovered that the flowers of two species of *Chrysanthemum* (sometimes known as *Tanacetum*), *C. cineranifolium* and *C. coccineum*, had strong insecticidal properties, and that their flowers could be dried, ground into powder and dusted on insects to kill them. First used in China, knowledge of them seems to have slowly spread west along the trade routes until it reached Europe, where the plants themselves came to be cultivated in the Balkans. The species came to be known as pyrethrum, and the active insecticide as pyrethrin; the insecticide ingredient was not easy to prepare, and in its early days, its unusual properties also led to it being kept a secret. After centuries of use it only became widely available in trade in the nineteenth century, although it still remained expensive. Since then, it has been used both as an (expensive) insecticide on crops and as a public health medication for dusting people to kill human body lice, which are important vectors of disease. It is an extremely potent insect neurotoxin that is harmful to people if ingested in large quantities, but is perfectly safe at low dosages on crops. It also breaks down rapidly when exposed to air and sunlight, does not accumulate in the environment, and does not cause any significant harm to wildlife. At the present time, it is still widely used in organic farming.

Arsenic Insecticides

The first known use of arsenic compounds was as medicine in ancient China, and at some point, the knowledge that these substances were biologically potent also led to their use as insecticides (see "Arsenic" in Chapter 5). These practices, however, do not seem to have reached the West, and it was only in the nineteenth century that arsenic started being used there for pest control. In the United States, for example, compounds like lead arsenate and copper acetate tri-arsenate came to be employed as sprays to kill caterpillars feeding on forest trees, apples and other crops. (The worst pest on apples in North America at the time was the codling moth, which was an introduced species from Europe.)

Soon, lead arsenate was being widely used against codling moths in apple orchards and on many other crops. As early as 1920, however, it was understood that it left toxic residues on fruit even after washing, although during that decade and the next, no effective substitute was identified, so arsenic compounds continued to be heavily used, even though poisonings and even some deaths among farm workers remained a regular occurrence. In the 1940s, the introduction of the first chlorinated hydrocarbon insecticide was welcomed not only for its effectiveness and low cost, but also because it was seen as a replacement for more hazardous arsenic-based pesticides. Chlorinated hydrocarbons quickly came to dominate the market, until it became apparent that they too could cause severe environmental damage. Eventually, other, less harmful, synthetic insecticides were developed which made it possible to produce high quality fruit without the use of either arsenic or chlorinated hydrocarbons. Even today, however, much commercially produced fruit contains measurable residues of possibly harmful pesticides.

Chlorinated Hydrocarbon Insecticides

By the late 1930s, chemistry's rapid progress during the early decades of the twentieth century had led to dramatic changes in the world of pesticides, most notably in the form of a new, synthetic compound developed by one of the world's oldest chemical companies, J. R. Geigy of Switzerland. Geigy had started in 1859 as a manufacturer of synthetic dyes, especially fuchsine (magenta), along with a wide variety of other chemicals. By the early twentieth century, one of its divisions had come to specialize in agricultural chemicals, including pesticides, which at the time were still dominated by arsenic compounds. In 1925 Geigy hired a young scientist named Paul Hermann Mueller, who had just completed his doctorate in chemistry at Basle University, and initially assigned him to work on the development of plant-based dyes and natural agents for tanning leather. Around 1935, however, Mueller redirected his research to search for new, synthetic compounds that could kill insects on contact, and in 1939 he showed that a chlorinated organic compound that had initially been synthesized in 1874 worked extremely well for this purpose. Chemically, it was called dichlorodiphenyl-trichloroethane, or 1,1,1-trichloro-2,2-bis(4-chlorophenyl)ethane—a mouthful of a name, which eventually became abbreviated to DDT.[60] As synthetic organic chemicals went, DDT was inexpensive and relatively simple to make. Early testing also suggested that it was much less toxic to people (and other mammals) than arsenic pesticides, while still amazingly effective at killing the insects that caused both human diseases and crop losses. Like pyrethrin, DDT was a neurotoxin which killed insects on contact.

In the same year that Mueller made his discovery, World War II erupted, and within two years the conflict had spread to engulf much of the world, severely disrupting international trade and cutting off easy access to the natural insecticide pyrethrin. Once DDT became commercially available in the early 1940s, it was therefore seized upon as filling a critical public health need, and throughout the rest of the war, it saved many lives (perhaps even millions), in particular by destroying malaria mosquitoes on Pacific islands where much fighting occurred, and also by sharply reducing deadly typhus through control of human lice on soldiers, refugees and other civilian populations. At the time, it seemed like a true miracle of chemical technology, and even earned Paul Mueller a Nobel Prize in Physiology and Medicine in 1948. (The Nobel Prize for Chemistry in that year went to the Swedish chemist Arne Tiselius, as Mueller's work was not thought to represent an important advance in the science of chemistry itself.)

Throughout the 1940s and 1950s, part of the enthusiasm that greeted the introduction of DDT stemmed from the belief that it was much less toxic than the commonly used arsenic-based insecticides that it had largely replaced. This was certainly true of its short-term effects on people, although it did not turn out to be the case for its overall long-term effects. After World War II, the manufacturers of DDT had sought successfully to extend its market to agriculture, and by 1950 it was being widely used on farms. Over roughly the next two decades—before its use in agriculture started to be restricted in the 1970s—an estimated 1.8 billion kilograms of DDT was produced and released into the environment.[61]

After truly amazing results in its early years of use, by the late 1940s DDT had also come to be less effective both in agriculture and in public health as the result of natural selection for resistance in pest species. In a number of places where large-scale agricultural crop spraying was carried out, for example, mosquitoes that carried malaria also began to become resistant.[62] One characteristic that initially made DDT so effective was its chemical stability. It was very slow to break down, and if it was sprayed on a field, it might still be actively killing insects days or weeks later. Yet this property also made it an even greater environmental menace. DDT and several of its breakdown products tended to stay around for years in the environment and also to accumulate in food chains so that animals higher up these webs of life concentrated larger and larger amounts of the pesticide along with its damaging breakdown products. DDT, and two of its chemical breakdown products, DDE and DDD, are extremely toxic to many animals, including fish and birds. In people, according to the USEPA, these three substances are also able to cause health problems over the long term by causing damage to the liver and reproductive system, while also increasing the likelihood of cancer.[63] Some authorities also believe that exposure to DDT and its breakdown products may increase people's risk of diabetes.

A number of countries still permit DDT use in disease control, and some still allow it in agriculture. As a result, it continues to be present in varying quantities almost everywhere, as global environmental cycles disperse water runoff and dust from agricultural fields.

Organophosphates

DDT was the first of a long series of synthetic chlorinated hydrocarbon insecticides that began to be produced during and after World War II, a number of which proved to be extremely harmful to human health and/or the natural environment. Around the same time that DDT was being developed in Switzerland, chemical researchers in neighboring Germany were discovering the first member of a different, entirely new class of synthetic insecticides, the organophosphates. These also attacked insect nervous systems (and often those of other animals), but disrupted them in a different way than did the organochlorines. To create this new class of compounds the German researchers combined different organic molecules with inorganic phosphoric acids; the first effective insecticide they made in this way was called parathion. After a number of years of use, however, parathion also proved to be quite harmful and was eventually banned in many places, while some other organophosphates that were less toxic continued to be used. We will have more to say about this group of insecticides in the next chapter when we examine the way that organophosphate insecticides damage animal nervous systems at the cellular level.

Serious questioning of DDT's safety began in the late 1950s after observers noted dramatic declines in bird populations in regions where heavy spraying had occurred. By the 1960s a movement aimed at environmental protection had begun to appear in industrialized countries, and one of its leading voices was the American Rachel Louise Carson (1907–1964). Both a highly trained biologist and a tremendously gifted writer, Carson first became widely known for her classic mid-twentieth-century book, *The Sea Around Us* (1951), which introduced marine biology and oceanography to a broad public.

By the late 1950s, free from financial pressures thanks to the success of *The Sea Around Us*, Carson had begun to focus her research and writing on the environmental impact of the widespread use of synthetic chemical pesticides. In 1962, she published a well-researched book entitled *Silent Spring* describing the widespread use of synthetic organochlorine and organophosphate pesticides in the United States, and the devastating effects this was having on natural communities. As with *The Sea Around Us*, *Silent Spring* quickly became a best-seller as well as a rallying point for environmental protection. In it Carson carefully documented the broadcast use of synthetic insecticides over the previous two decades along with the growing body of research describing the resulting damage to the environment and to human health. She also described and analyzed the post-World War II mindset in which many believed that all-powerful technology—think of the atomic bombs that had ended the war—might allow people to control everything to the point where some even dreamed of wiping problem insect species off the face of the earth. These attitudes had obviously been important to wartime military planners, but they had also been taken up by civilian leaders across industry and government in the United States and a number of other countries. This had led to cozy relationships in which chemical companies, government agricultural officials, and the political leaders to whom they reported, joined together to support pesticide campaigns like those in the United States aimed at eradicating the gypsy moth and the fire ant. These programs, though biologically doomed to failure from the start, produced huge and profitable orders for chemical companies and also seem to have given politicians and government officials a feeling that they were powerful and in control. Unfortunately, they also ended up causing significant damage to natural systems and to human health.

In the United States, one response to the concerns raised by Carson and others was the creation in 1970 of the United States Environmental Protection Agency (USEPA), which eventually grew to be a leading regulator of potentially hazardous synthetic chemicals. Another response was the creation in 1972 of the United Nations Environmental Programme (UNEP), which was intended "To provide leadership and encourage partnership in caring for the environment by inspiring, informing, and enabling nations and peoples to improve their quality of life without compromising that of future generations."[64]

Since its creation, the USEPA has worked to limit hazardous chemical use and to ban compounds which are particularly toxic or environmentally damaging. On an international level, the UNEP has provided leadership in the control of environmentally damaging synthetic chemicals that are known as *persistent organic pollutants*. Starting in 1995, the UNEP undertook to coordinate development of an international treaty to restrict or ban the use of the most harmful of these substances, focusing originally on a dozen types of chlorinated hydrocarbons: aldrin, chlordane, dieldrin, endrin, heptachlor, hexachlorobenzene, mirex, toxaphene, PCBs, DDT, dioxins, and polychlorinated di-benzo furans, most of which were pesticides, industrial intermediates or their by-products. Updates to the UNEP treaty, frequently referred to as the *Stockholm Convention*, have since banned or restricted use of additional organochlorines as well as several organic bromine and fluorine chemicals.[65] Not all countries have ratified the treaty, and some of these extremely toxic substances are still being used, and perhaps even still being manufactured. Due to their persistence and heavy past use, all are still present in the environment and will be for years to come. But at least the very worst offenders have been branded as such, and their use banned or severely limited in many countries.

Agricultural Productivity

The production of basic food is one of the most important, if less celebrated, themes of human history. Over more than 10,000 years, human populations continuously, if unevenly, grew as agriculture slowly expanded and more and more of earth's natural plant communities were displaced (see Chapter 3). Ever-increasing population leading to more and more *cleared* land has been the long-term, global pattern. In many local places, however, years of good crops and adequate food supplies commonly alternated with years of drought, frost, insect upswings or soil erosion to produce food shortages, famines and even outright starvation. In China alone, a significant famine was recorded in one or another locale nearly every year until the middle of the twentieth century, a record stretching back roughly 2,000 years.[66]

These tragic and bitter experiences created a grim determination to maintain food production, and often this led to a short- to medium-term outlook that tended to ignore the long-term harm caused by some agricultural methods. Among the problematic practices were the repeated growing of the same crops in the same fields, the cultivation of steep slopes, the over-grazing of pastures and range land, and the excessive irrigation of land in dry climates leading to salt accumulation. Unfortunately, if perhaps understandably, these attitudes tended to persist even after inexpensive, manufactured agricultural chemicals were introduced during the twentieth century. As a result, these new fertilizers, insecticides, fungicides and herbicides were often over-used or applied in ways that damaged human health, the soil and the entire natural world.

Plant Nutrition

Agricultural fields are strikingly different from natural plants communities, not just in their elimination of most of the native plants and animals which originally lived there, but also in their rapid depletion of nutrients and frequent build-up of plant pests and diseases. In natural ecosystems, essential mineral nutrients are retained in many different places—in the bodies of animals, in the trunks, branches and roots of woody plants and grasses, in fallen leaves, twigs, buds and fruits that are slowly decomposing on the soil surface, and in the soil humus which stores nutrients derived from plant and animal remains and droppings. In order to grow, plants need a regular supply of carbon, hydrogen and oxygen, all three of which are generally available from carbon dioxide in the air and water in the soil. But they also need at least thirteen other nutrients which fall into three categories based on the quantities required. Boron, copper, iron, manganese, molybdenum, nickel and zinc are needed only in miniscule amounts and are called *trace nutrients* (note that arsenic, mercury and lead are not mentioned). Calcium, magnesium and sulfur are needed in greater, but not very large amounts and are known as *secondary nutrients.* Potassium, phosphorus and nitrogen are needed in fairly large amounts and are called *primary nutrients.*

In intact plant communities, very few nutrients are lost through leaching and soil erosion, and most losses are replaced by new nutrients being made available to roots through the weathering of rocks present in and beneath the soil. Natural plant communities also contain tens, hundreds—or in some tropical areas—thousands of different plant species, each with slightly different nutritional requirements. When many different species grow together, the overall pool of soil nutrients is extracted in a more balanced way and plant pests and diseases specific to individual plant species rarely rise to high levels. As we have seen, however, farm fields have a very different ecology. Food plants have been selected to pack the greatest content of nutrients into the seeds, leaves and fruits that are harvested for human food, and all of these nutrients are efficiently removed from the ecosystem and end up either in people, or in the human waste stream where they are generally unavailable for future plant growth. (At times in history, human waste has regularly been returned to farm fields in an attempt to recycle some of its nutrients, but this kind of *night soil* system presents dangerous risks of bacterial, viral and parasitic disease.)

In agriculture, just one or a very few kinds of plants are often grown repeatedly in a given field, extracting a similar profile of nutrients. Even when crop rotation is practiced, it is often the case that agricultural land tends to becomes less fertile and less productive over time. On very rich, deep soils in steppe, prairie and river flood plains, the huge store of nutrients that was originally present can sometimes maintain fertility for a long time. But in areas of thin, infertile or steeply sloping soils like those which cover much of the earth's surface, declining crop plant growth can occur even after just one or two growing seasons.

Fertilizers

Over thousands of years, farmers tried to cope with declining fertility through a number of traditional techniques, many of which are being revived today by growers seeking to reduce or eliminate manufactured chemical inputs to farmland. First they spread all available animal manure onto their fields as a combined humus and fertilizer; second, they rotated crops in the same field from season to season and year to year; third, they left their fields *fallow* or unplanted on a regular basis, plowing in the diverse non-food plants ("weeds") that grew up during the off years; and fourth—especially in more recent times—they intentionally planted legumes and other soil enriching *green manure* plants in a rotation, incorporating their residues into the soil at the end of the growing season. By the nineteenth century, however, many areas in Europe and Asia were experiencing shortages of agricultural land coupled with a growing population, and there was simply not enough space to produce an abundance of food in traditional ways. Manure-producing farm animals often required pastures, hayfields or forage crops, which did not produce grain or other plants for human consumption, and some schemes of crop rotation and green manure similarly required large areas of temporarily unproductive land. Some food could be imported, and the introduction of potatoes from South America had helped by allowing more food to be produced on each acre. But overall, food production was again coming up against natural limits, while prices were rising.

During the first half of the nineteenth century, scientists had established both that plants needed potassium, phosphorus and nitrogen in significant quantities, and that they could take up these nutrients whether they were supplied in the more natural forms of manure and potash or in the form of inorganic (mineral) compounds. (Whether organic and inorganic nutrient sources truly provide fully identical nutrition to plants remains a subject of debate.) These discoveries spurred the search for new chemical fertilizers, although producing them required access to large amounts of the right kinds of potassium, phosphorus and nitrogen compounds, most of which were in limited supply worldwide. Potassium had traditionally been added to farm and garden soils in the form of potash, a term that referred originally to plant ashes. In earlier times, when human populations were smaller and there were more intact forests, extra potash was often locally produced by burning wood—indirectly mining the potassium present in forest soils. By the nineteenth century, however, much of Europe had been deforested—deforestation had started thousands of years earlier around the Mediterranean. Tree potash could still be purchased from places like Canada which possessed extensive areas of forest, but it was relatively expensive.[67]

Besides potassium, crops also required significant amounts of phosphorus and nitrogen. Mineral deposits of calcium phosphate rocks, originally laid down under the ocean, could be found in a number of different countries, but the phosphorous in them was locked up chemically, so that simply adding ground phosphate rock to the soil did not provide a great deal of additional phosphorus to plants. This fact led researchers to try to treat phosphate rocks in different ways to make their nutrients more available to plants. We saw earlier that strong acids were a leading product of the early chemical industry in Britain, and it was their ready availability which soon led to an important breakthrough in the production of fertilizer. In the late 1830s, John Bennett Lawes, a rich English landowner and agricultural experimenter, treated phosphate rocks with sulfuric acid to produce a water-soluble phosphorus compound that plants could readily absorb. Lawes made this discovery at his estate in Rothamstead, England, a location that he later developed into a leading agricultural research institute. By 1842, he had also set up a factory in Deptford where he was the first to manufacture this new fertilizer, which came to be known as *superphosphate*.[68]

Nitrogen was the third nutrient that plants required in large amounts. It made up nearly four fifths of the earth's atmosphere, and so was available literally everywhere. But in the atmosphere, it was nearly all present as an unreactive gas that could not be directly used by plants. In contrast, most plant nitrogen came from the recycling of biological nitrogen derived from decomposed animals and plants. In addition, lightning discharges in the atmosphere produced small amounts of usable nitrogen compounds which washed down in rain and snow to increase soil supplies, while several kinds of bacteria also played roles in either adding or removing usable nitrogen from farmers' fields and natural plant communities.

In the mid-nineteenth century, nitrogen and phosphorus compounds were also still available in the form of seabird droppings (guano) deposited over thousands of years by colonies of cormorants, pelicans and boobies, all of which fed on fish. It made very effective fertilizer, although it was relatively expensive. Large quantities of such coastal deposits, for example, were mined in Peru between 1845 and 1870 providing a short-term boost to the economy. But after only a few decades, this valuable resource became exhausted as the birds could only replenish the guano at a tiny fraction of the rate at which it was being removed. (Besides fertilizer, guano was also exported because its content of nitrates made it important as a raw material for the making of gunpowder and other explosives.)

Synthetic Nitrogen Fertilizer

The depletion of the largest guano deposits in South America in the third quarter of the nineteenth century had once again led to shortages of nitrogen compounds for agricultural use, since deposits of nitrate minerals are relatively uncommon. Phosphate rocks could be mined in several places around the world and potassium could be obtained from imported plant ash if necessary. But nitrogen was hard to replace. Traditionally, much of it had come from animal dung, but manure supplies in rural areas were inadequate to meet fertility needs, while in cities, large quantities of horse manure were more a waste material to be disposed of than a resource that could efficiently be returned to the land. Earlier in the nineteenth century, laboratory experiments with electricity had shown that an electric spark could be used to drive a small-scale version of the chemical reaction that occurred naturally in the air around lightning bolts, combining nitrogen and oxygen to make nitric oxide. In 1903, in Norway, this technique actually began to be used to make nitrogen fertilizer in a factory close to a large hydroelectric plant that provided very inexpensive electricity. As an industrial process, however, electricity-based nitric oxide synthesis was extremely inefficient, and for this reason, the nitrate fertilizer it produced was prohibitively expensive.

Over the last decades of the nineteenth century, the overall range of laboratory tools available for chemical synthesis had continued to multiply, and researchers had begun to discover that by carrying out chemical reactions in sealed vessels under different combinations of pressure and temperature—often in the presence of catalysts—they could make compounds that were impossible to synthesize in other ways. This, in turn, opened up many new possibilities and raised hopes that perhaps better ways could be found to produce chemical fertilizer, a pressing global need given the frequent shortages of food and recurring instances of famine. One possible approach was to try to synthesize the simple, natural, nitrogen-containing compound, ammonia, as a starting point for the making of nitrate—each ammonia molecule contained a single atom of nitrogen bonded to three of hydrogen. The hydrogen necessary for this synthesis could be obtained by passing an electric current through water—bubbles of hydrogen would be produced at one electrode while bubbles of oxygen appeared at the other. The nitrogen itself was of course available from the air, although for a long time there was no effective way to separate pure nitrogen from the atmosphere. Then, in 1883, the Polish physicists Zygmunt Wroblewski and Carol Olszewski succeeded in liquefying a small amount of nitrogen from the air, and by 1895 industrial-scale nitrogen production had become possible.

By the end of the nineteenth century, both the hydrogen and the nitrogen needed for ammonia synthesis were available in large quantities. Yet despite tremendous effort, scientists were still not able to find an efficient way to react the two together to make ammonia. Finally, in the first decade of the twentieth century, after a long series of careful experiments trying different combinations of pressure and temperature along with a variety of catalysts, the German physical chemist Fritz Haber developed a workable reaction. Well aware of the practical importance of his discovery, in 1909 he proposed to the giant German chemical company BASF that they collaborate on scaling up his laboratory process to allow industrial production of ammonia. Impressed by Haber's demonstrations, BASF agreed. It then spent four years under the leadership of Carl Bosch designing and perfecting the huge, high-pressure reaction vessels that were needed for continuous flow, mass production. BASF also developed inexpensive iron catalysts to use in the process along with a method for utilizing super-hot, high-pressure natural gas as the hydrogen source. Once this had been done, ammonia became one of the chemical industry's most important products, a position that it still retains today.[69] In the early twenty-first century, for example, more than 100 million metric tonnes were manufactured annually.[70] Most of it is used for fertilizer, although some also serves as raw material for making a wide range of other useful chemicals, particularly nitric acid, which is an essential starting material in the manufacture not only of fertilizers and explosives, but also of many other products.

In the twenty-first century, in fact, ammonia is being manufactured on such a large scale that its synthesis currently consumes more than three percent of all natural gas supplies, a footprint equal to more than one percent of the entire human energy supply. To produce the final fertilizer, ammonia is most often reacted with nitric acid to produce ammonium nitrate, a salt that can be readily blended into solid potassium–nitrogen–phosphorus fertilizer. It can also be combined with carbon dioxide to make urea nitrogen fertilizer. And sometimes, straight liquid ammonia under pressure is injected directly into the soil as a source of nitrogen, a practice particularly useful for growing corn (maize).[71]

In 1918, Fritz Haber was awarded the Nobel Prize for Chemistry for his ammonia synthesis, while Carl Bosch received the same award in 1931 for the development of industrial-scale ammonia production. In their honor, industrial ammonia synthesis came to be known as the Haber–Bosch process. But it should also be noted that Haber, who was an extremely zealous German patriot, also felt that it was his duty to use his scientific expertise to aid the German effort in World War I. Being a chemist, one idea he promoted was the use of poison gas as a weapon. In 1914, not long after the outbreak of war, France had tried using irritating chemicals in hand grenades, although they had proved ineffective on the battlefield. On the German side, Haber soon realized the possibility of using poisonous chlorine gas as a weapon. The German chemical industry was already producing large quantities of it as a by-product of chemical manufacturing, and it was known to be corrosive and potentially deadly if inhaled. In 1915, after trench warfare had settled down to a kind of stalemate, Germany decided to try out this new weapon, and Haber personally traveled to the battlefield to supervise its first use at the battle of Ypres in April 1915. Militarily, it did not prove to be decisive that day, but it was the first time that deadly, factory-made chemicals had intentionally been released on the battlefield. For the rest of his life, Haber was widely criticized by other scientists for his involvement, with some even referring to him as the *father of poison gas*.[72]

On balance, it is hard not to view the development of ammonia synthesis as progress, even though its application to explosives probably meant an increase in the manufacture and use of munitions, and almost certainly contributed to the duration and destructiveness of both world wars. By some estimates, 40 percent of the current human food supply would disappear without the fertilizer produced through industrial ammonia synthesis, and this could mean starvation for hundreds of millions. Synthetic nitrogen fertilizer was also essential to the *green revolution* of 1960–1980, which dramatically reduced world hunger and deaths from famine despite a rapidly increasing human population (see Chapter 3). The green revolution was a set of international programs in agriculture that allowed crop yields to more than double in many developing nations by combining genetically improved seed with increased amounts of fertilizers, pesticides and irrigation water. It saved many lives, although many believe that it also exacted a significant environmental toll.

Environmental Impacts of Chemical Fertilizers

As has often been the case with innovative technologies, the full cost of inexpensive fertilizer manufacture has only become clear over time as the resulting environmental damage has mounted up. Unregulated application of low-cost nitrogen fertilizer has frequently led to over-fertilization. Faced with unpredictable weather, uncertain rates of crop growth, and a strong motivation to earn a good living, farmers—often with little or no biological or chemical education—have frequently ended up adding more nitrogen to their fields than their crops are actually able to use in a given growing season. This excess, along with any extra phosphorus and potassium, invariably ends up washing out of fields into ponds, lakes, rivers and streams, ecosystems whose animals and plants are naturally adapted to the presence of very low levels of these nutrients.

Normally, small aquatic plants and plant-like microorganisms exist at modest population sizes supported by the low nutrient levels present in most fresh waters. In such habitats, the overall growth of plants is limited by plant-eaters, and a complex web of life can be sustained. But once artificial fertilization of fields with water-soluble chemical fertilizers begins, natural aquatic systems are often disrupted because excess nitrogen and phosphorus cause explosive growth of water plants and microbes to the point where they overwhelm the natural cycles in the water bodies. When this happens, dissolved oxygen levels in the water drop sharply, initially because the exploding population of plants use up oxygen at night, even though their photosynthesis also generates some oxygen during the day. As oxygen is depleted, the crowded plants begin to die back, with dead plants being decomposed by masses of bacteria whose metabolism further reduces dissolved oxygen levels. Under such conditions, fish and many other animals quickly die, while birds and mammals dependent on the fish also disappear. In the first half of the twentieth century, these effects were usually seen only in water bodies quite close to the most heavily fertilized fields. But as fertilizer use enormously expanded, such effects also began to be seen in entire river systems around the world.[73]

One large river basin heavily impacted in this way was the Missouri–Mississippi River system in North America which drains an area of more than one million square miles (more than two and a half million square kilometers) including much of the United States' most productive farm land. In the twentieth century, discharges of farm and industrial chemicals began to cause major changes to the Missouri–Mississippi. Waste from industry is most prominent in the lower Mississippi Valley, especially in Louisiana where a large chemical industry grew up to take advantage of nearby Gulf Coast oil and gas deposits. Runoff of agricultural fertilizer, insecticides and herbicides occurred over a much wider area in countless locations where intensive agriculture was carried out.

Being a single drainage basin, runoff from thirty-one different American states finds its way into the Mississippi through its many tributaries. After World War II, North American agriculture became increasingly intensive, incorporating many of the techniques which later came to be exported to developing nations as part of the green revolution. These included planting more productive crop varieties developed through scientific plant breeding, heavier use of machinery, and increasing reliance on fertilizers, pesticides and herbicides, along with large-scale, enclosed factory farming of livestock. The land itself was changed too: in many states within the Mississippi drainage, large areas of seasonally damp or wet acreage were permanently dried out by the installation of sub-surface drainage pipes which also had the effect of increasing the rate at which excess farm chemicals drained off fields and entered water bodies.

Taken together, these changes increased the runoff of water soluble nitrogen and phosphorus within the Mississippi's huge watershed. In the second half of the twentieth century growers dramatically increased their overall fertilizer use in order to raise yields per acre; factory farms produced so much concentrated animal wastes, especially from pigs and chickens, that it became difficult to keep nutrients from leaching into surrounding waterways; so much low-lying acreage was cultivated that excess fertilizer runoff tended to quickly make its way into river drainages before nutrients could be extracted by natural plant communities. The main stem of the Mississippi River thus became a concentration point for vast amounts of agricultural runoff, just as the Rhine had earlier been for European industrial discharges. The Mississippi itself had long since lost a great deal of its original ecosystem and community of living things. But now, as it flowed slowly down to the Gulf of Mexico, it brought with it so much dissolved phosphorus and nitrogen that its outflow even began to cause serious damage to the ocean. Water pollution is easy to understand when one sees a small pond covered with a suffocating layer of green plants and bacteria—the pond looks as if it is being smothered, and it is not hard to imagine that there is little or no oxygen in its waters. But in a large oceanic area like the Gulf of Mexico, it is harder to imagine that river-borne pollution could cause truly significant damage. The unfortunate truth is that it can.

Beginning in the 1970s, environmental scientists started to notice an extensive *dead zone* along the coast of Louisiana and Texas caused primarily by the Mississippi's nutrient laden discharge. Excess agricultural—or in some cases industrial—phosphorous and nitrogen carried along by the river over-stimulated the growth of algae in the Gulf, leading to similar conditions in coastal waters as had often occurred in shallow scummed-over ponds. Almost all oxygen was used up, and as a result, fish, shrimp and other marine animals died off. In 2008, for example, the dead zone in the Gulf of Mexico produced by pollution from the Mississippi River covered 8,000 square miles (20,720 square kilometers), and it was only one of many such areas around the globe.[74] Almost 150 dead zones have been documented in the world's oceans, most related to river and coastal discharges of fertilizer runoff, animal waste and sewage discharge.[75] Artificial fertilizers proved to be a great boon to human food production over the last century, but as with so many other technological advances, they also caused a great deal of unexpected, unintended damage to ancient life-support systems which are critical for long-term human survival.

Modern chemistry and chemical manufacturing have played essential roles in creating the modern world; they have contributed thousands of new materials and products that have significantly enhanced our standard of living. Many manufactured substances have also been essential to the development of other technologies in transportation, industry, communications, electric power, medicine, and other fields. This great success, however, has not come without significant costs in the form of damage to earth's living environment and to human health. We will explore these topics further in the next chapter.

Notes

1   R. Shahack-Gross et al., 2014, Evidence for the repeated use of a central hearth at Middle Pleistocene (300 ky ago) Qesem Cave, Israel, *Journal of Archaeological Science* 44: 12–21.

2   Wikipedia, Venus of Dolni Vestonice, https://en.wikipedia.org/wiki/Venus_of_ Doln%C3%AD_V%C4%9Bstonice.

3   H. Watzman, Earliest evidence for pottery making found, *Nature* (June 1), www.nature.com/news/2009/090601/full/news.2009.534.html.

4   Wikipedia, Catalhoyuk, https://en.wikipedia.org/wiki/%C3%87atalh%C3%B6y% C3%BCk.

5   Internet Library of Serb Culture, Archaeology: Vinca, http://rastko.rs/arheologija/ vinca/vinca_eng.html.

6   A. Cotterell and R. Storm, 1999, *The Ultimate Encyclopedia of Mythology*, Hermes House.

7   K. Killgrove, 2012, Lead poisoning in Rome—the skeletal evidence, www.powered- byosteons.org/2012/01/lead-poisoning-in-rome-skeletal.html; M. A. Lessler, 1988, Lead and lead poisoning from antiquity to modern times, *Ohio Journal of Science* 88(3): 78–84.

8   National Mining Association, History of gold, www.nma.org/pdf/gold/ gold_history.pdf.

9   M. Hunter, 2004, Robert Boyle, *Oxford Dictionary of National Biography*, www.oxforddnb.com.

10 J. P. Poirier and R. Balinski, 1996, *Lavoisier, Chemist, Biologist, Economist*, University of Pennsylvania Press.

11 B. Warner, 2013, Ancient Roman concrete is about to revolutionize modern architecture, *Bloomberg Business Week* (June 14), www.businessweek.com/articles/2013- 06-14/ancient-roman-concrete-is-about-to-revolutionize-modern-architecture.

12 D. M. Kiefer, 2001, Sulfuric acid: pumping up the volume, *Today's Chemist At Work* 10(9): 57–58, http://pubs.acs.org/subscribe/journals/tcaw/10/i09/html/09chemch.html.

13 F. Aftalion, 1991, *A History of the International Chemical Industry*, 2nd edition, University of Pennsylvania Press.

14 Your Dictionary, Nicolas Leblanc facts, http://biography.yourdictionary.com/nicolas- leblanc.

15 Wikipedia, Leblanc process, https://en.wikipedia.org/wiki/Leblanc_process.

16 Halton Borough Council, Contaminated Land Strategy 2008–2013, www3.halton.gov.uk/Pages/planning/policyguidance/pdf/evidencebase/Viability/ Halton_Contaminated_Land_Strategy_2008-2013.pdf; ChemEurope.com, Leblanc process, www.chemeurope.com/en/encyclopedia/Leblanc_process.html.

17 Wikipedia, Leblanc process, https://en.wikipedia.org/wiki/Leblanc_process.

18 Wikipedia, Alkali Act 1863, https://en.wikipedia.org/wiki/Alkali_Act_1863.

19 NNDB, Marcellin Berthelot, www.nndb.com/people/892/000100592.

20 G. Lunge, 1887, *Coal Tar and Ammonia*, Taylor & Francis, https://archive.org/stream/ coaltarandammon00lunggoog#page/n8/mode/2up; A. Bosch, 2010, Toxic coal-tar

cleanup to cost New York $3 billion, *Times Herald Record* (February 7), www.recordonline.com/article/20100207/News/2070337?template=printart.

21 L. M. Pinsker, 2005, Feuding over the origins of fossil fuels, *Geotimes*, www.geotimes.org/oct05/feature_abiogenicoil.html; USGS, Energy Resource Program, Organic origins of petroleum, http://energy.usgs.gov/Geochemistry Geophysics/GeochemistryResearch/OrganicOriginsofPetroleum.aspx.

22 A. S. Travis, 2004, Sir William Henry Perkin, *Oxford Dictionary of National Biography*, www.oxforddnb.com.

23 R. Brightman, 1956, Perkin and the dyestuffs industry in Britain, *Nature* 177: 815–821.

24 E. Homburg, A. S. Travis and H. G. Schroeter (eds.), 1998, *The Chemical Industry in Europe, 1850–1914: Industrial Growth, Pollution, Professionalization*, Kluwer Academic; B. Ross and S. Amter, 1993, *The Polluters: The Making of Our Chemically Altered Environment*, Oxford University Press; A. S. Travis, 1993, *The Rainbow Makers: The Origins of the Synthetic Dyestuffs Industry in Western Europe*, Lehigh University Press.

25 E. Homburg, A. S. Travis and H. G. Schroeter (eds.), 1998, Pollution and the chemical industry—the case of the German dye industry, in *The Chemical Industry in Europe, 1850–1914: Industrial Growth, Pollution, Professionalization*, Kluwer Academic, pp. 183–200; New Jersey Hazardous Substance Fact Sheet, 1-napthylamine, http://nj.gov/health/eoh/rtkweb/documents/fs/1325.pdf.

26 American Chemical Society, CAS Registry, www.cas.org/expertise/cascontent/registry/regsys.html#q1.

27 Scorecard, Chemical profiles, http://scorecard.goodguide.com/chemical-profiles/def/universe.html; Scorecard, High production volume chemicals, http://scorecard.goodguide.com/chemical-profiles/def/hpv.html.

28 R. A. Relyea, 2009, A cocktail of contaminants: how mixtures of pesticides at low concentrations affect aquatic communities, *Oecologia* 159(2): 363–376.

29 Wikipedia, Ignacy Lukasiewicz, https://en.wikipedia.org/wiki/Ignacy_%C5%81 ukasiewicz.

30 ExxonMobil, ExxonMobil chemicals: petrochemicals since 1886, http://exxonmobil.com/Benelux-English/about_who_history_chemical.aspx; Wikipedia, Petrochemical, https://en.wikipedia.org/wiki/Petrochemical.

31 The Encyclopedia of Earth, Petroleum refining, www.eoearth.org/view/article/155206.

32 Wikipedia, Oil refinery, https://en.wikipedia.org/wiki/Oil_refinery.

33 Wikipedia, Plastic, https://en.wikipedia.org/wiki/Plastic; Encyclopedia.com, Polyester, www.encyclopedia.com/topic/polyester.aspx.

34 Wikipedia, Resin identification code, https://en.wikipedia.org/wiki/Resin_identification_code.

35 NOAA, National Ocean Service, The great Pacific garbage patch, http://oceanservice.noaa.gov/podcast/june14/mw126-garbagepatch.html; Wikipedia, Great Pacific garbage patch, https://en.wikipedia.org/wiki/Great_Pacific_garbage_patch.

36 Wikipedia, Great Atlantic garbage patch, https://en.wikipedia.org/wiki/North_Atlantic_garbage_patch.

37 Wikipedia, Periodic table, https://en.wikipedia.org/wiki/Periodic_table.

38 Granta Material Intelligence, Annual world production, http://inventor.grantadesign.com/en/notes/science/material/C04%20Annual%20world%20production.htm.

39 US National Cancer Institute, Vinyl chloride, www.cancer.gov/about-cancer/causes-prevention/risk/substances/vinyl-chloride.

40 European Council of Vinyl Manufacturers, Lead stabilizers, www.pvc.org/en/p/lead-stabilisers.

41 World Health Organization, Dioxins and their effects on human health, www.who.int/mediacentre/factsheets/fs225/en.

42 Insititute of Medicine, 2004, *Veterans and Agent Orange: Length of Presumptive Period for Association between Exposure and Respiratory Cancer*, National Academies Press, www.nap.edu/read/10933/chapter/1.

43 World Health Organization, Dioxins and their effects on human health, www.who.int/mediacentre/factsheets/fs225/en.

44 USEPA, Learn about dioxin: environmental laws that apply to dioxin, www.epa.gov/dioxin/learn-about-dioxin#tab-2.

45 USEPA, An inventory of sources and environmental releases of dioxin-like compounds in the US for the years 1987, 1995, and 2000 (final, Nov 2006), https://cfpub.epa.gov/ncea/dioxin/recordisplay.cfm?deid=159286; Wikipedia, Dioxins and dioxin-like compounds, https://en.wikipedia.org/wiki/Dioxins_and_dioxin-like_compounds.

46 European Commission, 2010, On the implementation of the Community Strategy for dioxins, furans, and polychlorinated biphenyls, COM (2001)593, Third progress report, http://eur-lex.europa.eu/legal-content/EN/TXT/PDF/?uri=CELEX:52010DC0562&from=EN.

47 Stockholm Convention, PCBs Elimination Network (PEN), http://chm.pops.int/Implementation/PCBs/PEN/tabid/438/Default.aspx.

48 C. K. Drinker et al., 1937, The problem of possible systemic effects from certain chlorinated hydrocarbons, *Journal of Industrial Toxicology and Hygiene* 19: 283–311.

49 Agency for Toxic Substances and Disease Registry, Toxicological profile for polychlorinated biphenyls, www.atsdr.cdc.gov/ToxProfiles/TP.asp?id=142&tid=26; K. Breivik et al., 2002, Towards a global historical emission inventory for selected PCB congeners—a mass balance approach: 1. Global production and consumption, *The Science of the Total Environment* 290(1–3): 181–198.

50 USEPA, EPA Cleanups: GE-Pittsfield/Housatonic River Site, www.epa.gov/ge-housatonic.

51 USDA, 2007, Table 45, Fertilizers and chemicals: 2007 and 2002, 2007 Census of agriculture—United States data, www.agcensus.usda.gov/Publications/2007/Full_Report/Volume_1,_Chapter_1_US/st99_1_044_045.pdf.

52 Extoxnet, 2,-4 D, http://pmep.cce.cornell.edu/profiles/extoxnet/24d-captan/24d-ext.html.

53 J. Perera and A. Thomas, 1985, This horrible natural experiment, *New Scientist* (April 18): 34–36.

54 Wikipedia, Dioxins and dioxin-like compounds. https://en.wikipedia.org/wiki/Dioxins_and_dioxin-like_compounds.

55 US Department of Veterans Affairs, Veterans' diseases associated with Agent Orange, www.publichealth.va.gov/exposures/agentorange/diseases.asp.

56 Wikipedia, Atrazine, https://en.wikipedia.org/wiki/Atrazine.

57 Ibid.; Toxipedia, Atrazine regulation in Europe and the United States, http://toxipedia.org/display/toxipedia/Atrazine+Regulation+in+Europe+and+the+United+States.

58 D. J. Futuyma and A. A. Agrawal, 2009, Macroevolution and the biological diversity of plants and herbivores, *PNAS* 106(43): 18,054–18,061.

59 Wikipedia, Pesticide, https://en.wikipedia.org/wiki/Pesticide.

60 Nobelprize.org, Paul Mueller biography, http://nobelprize.org/nobel_prizes/medicine/laureates/1948/muller-bio.html.

61 Wikipedia, DDT, https://en.wikipedia.org/wiki/DDT.

62 Ibid.

63 Extoxnet, DDT, http://pmep.cce.cornell.edu/profiles/extoxnet/carbaryl-dicrotophos/ddt-ext.html.

64 United Nations Environment Programme, Mission, www.unep.org/Documents.Multilingual/Default.asp?DocumentID=43.

65 Australian Government, Stockholm Convention on Persistent Organic Pollutants, www.environment.gov.au/protection/chemicals-management/pops.

66 Wikipedia, List of famines, https://en.wikipedia.org/wiki/List_of_famines.
67 Wikipedia, Potash, https://en.wikipedia.org/wiki/Potash.
68 F. M. L. Thompson, 2004, Sir John Bennet Lawes, *Oxford Dictionary of National Biography*, www.oxforddnb.com.
69 Nobelprize.org, Nobel Prize in Chemistry 1918, Fritz Haber: award ceremony speech, www.nobelprize.org/nobel_prizes/chemistry/laureates/1918/press.html.
70 Wikipedia, Ammonia, https://en.wikipedia.org/wiki/Ammonia.
71 Ibid.
72 Smithsonian.com, Fritz Haber's experiments in life and death, http://blogs.smithsonianmag.com/history/2012/06/fritz-habers-experiments-in-life-and-death.
73 Encyclopedia of Earth, Eutrophication, www.eoearth.org/view/article/51cbedc37896bb431f693ba8.
74 J. A. Achenbach, 2008, "Dead zone" in the Gulf of Mexico, *Washington Post* (July 31), www.washingtonpost.com/wp-dyn/content/story/2008/07/31/ST2008073100349.html; Goddard Earth Sciences Data and Information Services Center, Dead zones, http://disc.sci.gsfc.nasa.gov/education-and-outreach/additional/ science-focus/ocean-color/science_focus.shtml/dead_zones.shtml.
75 Wikipedia, Dead zone (ecology), https://en.wikipedia.org/wiki/Dead_zone_%28ecology%29.

5
CHEMICALS IN THE ENVIRONMENT

Mechanisms of Chemical Toxicity and Carcinogenesis

If one is to understand what it means for something to be toxic or cancer-causing, it is important to begin by considering the intricate structures and workings of cells, for that is where most toxins, poisons and carcinogens do their damage. In some ways this fact is very unfortunate because cells' great structural and biochemical complexity—the result of more than three billion years of evolution and diversification—throw up a kind of barrier to popular understanding about the harm that can be done to living things by chemical pollution and radiation. As an introduction to cellular toxicity and the risk of cancer, we will look at four different kinds of toxic effects: specific substances that can poison the nervous system and cause paralysis and death; molecules that cause harm by mimicking hormones; chemical elements such as lead and arsenic which can produce a wide range of toxic effects in many different parts of the body; and substances that can increase the risks of cancer by damaging cells' genetic coding—the underlying biochemical recipes critical for normal functioning.

Animals with complex bodies first appeared on earth between five and six hundred million years ago, and an important part of their evolution was the development of two key kinds of internal control mechanisms: nervous systems and hormonal systems. The former arose as a way of controlling voluntary and involuntary bodily movements; of allowing animals to respond quickly to threats and opportunities in the environment; and of carrying out essential internal functions. Nervous systems—and the sense organs that developed with them—made it possible for animals to avoid predators and to better pursue food, mates and shelter. They also enhanced the functioning of critical body systems by creating rhythmic contractions in respiratory, circulatory and digestive organs. Hormonal systems, in contrast, evolved as chemical signaling systems designed to coordinate the complex growth, development and functioning of multi-celled creatures which possessed not only enormous numbers of cells, but also specialized organs and body parts. Hormonal systems typically employ very small quantities of biologically active molecules to coordinate essential body processes over extended periods of time; they commonly help to regulate functions like growth, sex determination, reproduction, water balance, digestion and metabolism.

People first became able to observe cells in the mid-1600s following the invention of the earliest microscopes. After that, it took roughly 200 years for scientists to begin to appreciate that cells really made up the essence of all living things. As late as the middle of the nineteenth century, cells were still often thought of as little structural units—a bit like bricks—infused with some mysterious *vital essence* that allowed them to carry out the surprising and impressive functions of life. Before the end of the nineteenth century, however, science became able to probe more deeply into the underlying structure and function of cellular life. This became possible because advances in physics made possible dramatically improved microscopes; new synthetic cellular stains were developed; and some of the fundamental principles of biochemistry began to be understood. From the 1870s on throughout the rest of the nineteenth and twentieth centuries, a vast empire of cellular structures, molecules and processes was slowly revealed by the diligent work of thousands of researchers around the world. Comparing them to explorers, it would be fair to say that what they discovered amounted not just to the description of new continents, but of entire new planets, so complex, varied and unexpected were the structures and dynamic processes that were discovered inside cells.

Nerve cells, or neurons, offer one example of this. First seen clearly in the late 1870s after the development of improved staining techniques, their workings only began to be understood in the early twentieth century after researchers discovered that they worked in part by triggering tiny electrical impulses along their long, skinny, thread-like cell projections. Functioning like wires that were much finer than a human hair, these structures linked one area of the body to another. These projections, called *axons*, could be extraordinarily long, sometimes as long as several feet, although still not generally long enough to reach all the way from limbs or organs to the brain and back. Instead, series of such cells, stretched end to end in a relay-race kind of fashion, carried out long-distance signaling in the bodies of animals. Once this was understood, researchers began to focus on the mechanisms by which the linked cells in an individual nervous pathway communicated with one another. It had been known throughout the nineteenth century that electricity had something to do with nerve signaling, but by the turn of the twentieth century some scientists had also begun to suspect that chemicals substances—*neurotransmitters*—might also be critical to the conduction of nerve impulses from one nerve cell to another. The first neurotransmitter identified, acetylcholine, was discovered around 1920, and throughout the rest of the twentieth century, dozens of additional chemical substances were identified that also played roles as chemical communicators in animal nervous systems.

In the nineteenth and the early twentieth century, one serious barrier facing scientists studying the workings of nerve cells was the inability of light microscopes to reveal the details of the tiny regions where adjacent nerve cells came into contact. In the 1930s, however, physicists in Germany developed the first workable electron microscope, a device that eventually made it possible to clearly visualize cell details that were far too small to be seen with light microscopes. By 1945, scientists had overcome a variety of engineering challenges and had actually begun using the new devices to examine cells. Capable of revealing details hundreds of times smaller than those which could otherwise be seen, the new instruments soon revealed an astonishing array of intricate structures, most of which were located inside cells. But that was not all they showed. The electron microscope also revealed remarkable details of synapses, the junctions where one nerve cell contacted another, as well as of motor end plates, the structures where neurons came into contact with muscles in order to control their movements. From that point on, progress accelerated, and the electron microscope revealed more and more of the nervous systems' finer structures, while experimental and biochemical studies shed additional light on how nerves and muscles actually function.

Most often, it turned out, two connected nerve cells came together in an elegantly structured junction called a chemical synapse which contained a very narrow space across which tiny amounts of a number of neurotransmitters—highly specialized chemicals—could rapidly move. In such cases, the continuation of a signal from one nerve fiber to another depended upon the rapid release of the right amount of one or more neurotransmitters into the synapse, an event adequate to stimulate the next nerve in the chain. Around the same time, analogous synapse-like structures were discovered joining the trigger sites on muscles (motor end plates) with the motor neurons that controlled their contraction. Over the course of the twentieth century, dozens of different substances were found to function as neurotransmitters in animals, and acetylcholine turned out to be one of the more common ones. It proved to be important not only in nerve synapses but also in the motor end plates of muscles. A complete understanding of its workings and significance, however, took decades to unravel, and some of the most important research that helped to clarify its function in the nervous system came from a surprising direction. (Less commonly, some nerve synapses were also found to function solely by electricity, being based on even narrower structures known as *gap junctions* which connected adjacent nerve cells so closely that electrical impulses could directly jump across them from one cell to another. Over hundreds of millions of years of animal evolution, however, electrical synapses had turned out to be less flexible than chemical ones, their one great advantage being that they could convey signals much more rapidly between adjacent cells than could chemical synapses. For this reason electrical synapses are mostly limited to rapid escape reflexes, especially in small invertebrates, where tiny intervals of time can spell the difference between life and death.)

Neurotoxins: Organophosphates

In 1936, a team of German chemists at the I. G. Farben chemical company was experimenting with a class of compounds called organophosphates in hopes of developing new types of insecticides. Instead, they discovered a substance that was remarkably poisonous to human beings, a molecule they named Tabun, and which they soon developed into a military nerve gas. At the time, the Nazi Government was aggressively pursuing armaments development in anticipation of World War II, and it redirected the research team's main focus to the development of chemical warfare agents. Soon, they managed to synthesize additional nerve gases including Sarin and Soman, along with Parathion, the first usable organophosphate insecticide. Each one of these agents caused paralysis and death in animals in varying degrees, and even though the exact mechanism of toxicity was not fully understood at the time, these discoveries stimulated a great deal of new research, some of it military work designed to discovery possible antidotes to nerve gases.

Normally, when a grasshopper wants to hop, or a soldier wants to walk or fire a gun, their brains send signals along trains of neurons until they reach muscles in the limbs which contract to cause movement. One essential part of this process is the controlled release of acetylcholine into synapses along the route of the train of nerves that connect to the motor end plates of the muscles. Under ordinary conditions, the acetylcholine released by the stimulated nerves is present just long enough to trigger the downstream nerve cell or muscle to *fire* before the acetylcholine is quickly broken down and recycled. This, in turn, allows the synapse to return to the state it was in before the signal crossed, ready to react appropriately to the next command issuing from the central nervous system. The neurotransmitter acetylcholine evolved through the combination of two compounds that were present early on in many animals, choline and acetyl-CoA. As animals evolved more complex nervous systems, acetylcholine was one of the chemical messengers that was recruited to function as a signaling chemical in synapses and muscle end plates. One of the things that made it suitable was the fact that it was both readily synthesized and readily broken down to its original (inactive) starting materials. The inactive parts—choline and acetyl-CoA—could easily be combined to make the neurotransmitter, while the finished acetylcholine could easily be broken down again to choline and acetyl-CoA once its signal had been received. This quality made it function like an inexpensive, fast-acting switch that nerve cells could use to stimulate other cells. Of course, in order for it to work, there also had to be a mechanism for breaking down the acetylcholine, and in the 1930s, a protein called *acetylcholinesterase* was discovered which proved to function as the *off* switch.

Acetylcholinesterase turned out to be a very efficient enzyme, with each molecule able to break down 25,000 molecules of acetylcholine per second; working out its function made possible the first real understanding of the poisonous properties of both organophosphate insecticides and nerve gases.[1] Surprisingly enough, it turned out that these toxins did not interfere at all with the manufacture of acetylcholine or with its movement across the synapse or muscle end plate. Instead, they worked by irreversibly combining with acetylcholinesterase in a way that completely deactivated its function as an *off* switch. Once this happened, nerve synapses and motor end plates were unable to relax and could not regenerate their potential to transfer new signals. In essence, these pesticides and nerve gases permanently locked the synaptic switches into the *on* position. Muscles—including those associated with breathing—were no longer able to relax, and instead became permanently contracted, rapidly producing paralysis and death.

This is the basic mechanism underlying the toxic properties of a range of organophosphates, although minor differences in their chemical makeup significantly modify their degree of toxicity to humans and to different kinds of animals. Nerve gases like Sarin and VX are extremely deadly to people even at miniscule doses, while organophosphate pesticides like malathion and diazinon are not particularly toxic to people at the doses normally used in agriculture, although they are still lethal to many pest and non-pest insect species, as well as to some vertebrates. (A single ounce of the synthetic organophosphate nerve gas sarin, in contrast, is enough to kill more than 50,000 people.) Malathion, which is widely used in agriculture and mosquito control, is not itself a major environmental pollutant because of its rapid breakdown. Given sufficient exposure, however, it can still be a hazard for people because it is converted in the body to a far more toxic related substance called malaoxon, which *is* able to cause a wide range of serious neurological symptoms and even death. There is also some suggestive evidence from animal studies that it may increase the likelihood of certain cancers.[2]

The toxic effects of nerve gases and organophosphate insecticides which result from their inactivation of acetylcholinesterase in synapses provides a simple but clear illustration of one way that chemical substances can damage living cells by blocking a critical step in an essential cellular pathway. In this case, the agents deactivate a working molecule once it has been synthesized. In other cases, toxic substances can work by blocking the formation of the working molecule in the first place. And sometimes, interference results not from neutralizing or blocking the synthesis of the working molecule itself, but by instead blocking a *helper* substance that also must be present in order for the cellular mechanism to work properly.

Parathion, the original organophosphate insecticide developed by I. G. Farben in the 1940s, began to be used commercially after World War II, and its importance increased considerably in the 1950s and 1960s after evidence began to accumulate of the damage caused by organochlorine compounds such as DDT. Widespread use of parathion, however, soon led to more and more cases of poisoning and deaths in agricultural workers, and sometimes even in non-workers living nearby, along with evidence of mounting damage to wildlife especially fish, birds and bees. People poisoned by it experienced a range of symptoms related to its interference with normal nerve action, including headache, convulsions, vomiting, and finally loss of the ability to breathe. During the 1980s, 1990s and 2000s, parathion came to be banned entirely in more than a dozen countries, while more and more restrictions and conditions were placed on its use in places where it continued to be used.[3] In the United States, where previously it had been widely applied, its uses were first restricted to people using safety equipment, and then to workers licensed to apply it, preferably by aerial spraying. Looking back, it seems clear that use of a potent, broad brush neurotoxin like parathion on vast acreage of agricultural crops was an ill-conceived use of technology.

In recent decades, damage caused by organochlorine and organophosphate insecticides led to the search for new classes of compounds to use against insects. From the 1980s onwards attention shifted to synthetic analogues of the naturally occurring plant insecticide nicotine, a substance which had been extensively used before synthetic insecticides became available, but was then phased out due to its toxicity to mammals. Like nicotine itself, the new *neonicotinoids*—including the world's current leading synthetic insecticide imidacloprid—work by interfering with chemical synapses which join connecting nerves and nerves to muscles. They are much less toxic to mammals than nicotine itself but nonetheless have recently been implicated in dramatic declines in honeybee populations worldwide. In 2013 the European Union (EU) instituted a limited multi-year ban on several neonicotinoids including imidacloprid.[4] Their relative newness makes it hard to fully assess their long-term environmental impacts.

Endocrine Disruptors: DEHP and Bisphenol A

Organophosphates—both insecticides and nerve gases—specifically attack animal nervous systems. On the plus side, most of them break down fairly quickly in the environment. Generally, the symptoms they cause show up within minutes, hours or at most days. Substances which interact with and damage hormonal systems, in contrast, often take longer to show their effects and since many hormone-mimicking chemicals are more stable in the environment, their impacts can be felt over a longer period of time and even at considerable distance from the site of chemical use. In the late 1980s, for example, researchers studying fish and other animals showing signs of sexual and developmental problems discovered that the bodies of affected individuals often contained synthetic chemicals that seemed to interfere with critical body functions. Several of these molecules had structures that were similar enough to those of naturally occurring hormones to influence the production of critical substances like estrogen, testosterone and thyroxin. Synthetics like these came to be known as *hormonally active agents*, *endocrine disrupting compounds*, or simply *endocrine disruptors*. Most commonly, they affected the production and metabolism of sex hormones like estrogen. Such pollutants impacted all stages of the life cycle, but were particularly damaging to pregnant females, embryos, infants and children of humans and other vertebrates.[5]

DEHP

Once the phenomenon of endocrine disruption had been recognized, researchers began to look for common synthetics that were hormonally active in human and animal tissues, and one compound they soon identified was bis (2-ethylhexyl) phthalate (DEHP), the most common plasticizer used in making polyvinyl chloride plastic. They then tested it in experimental animals and found that it did have the potential to alter normal hormonal function. In some studies, young male laboratory animals exposed to it were unable to produce adequate amounts of male hormones like testosterone during the course of development. When this happened, it led to significant abnormalities of the penis, testes and other reproductive structures, which then translated into reduced fertility. Further research in humans produced some evidence that unusually heavy exposure to this widely used substance also interfered with the development of the penis and testicles in boys, and led to lower sperm counts in men. In response to these findings, in 1999 the EU instituted a ban on all use of phthalates in plastics destined for children's toys, and similar restrictions have since been put in place in the United States and Australia.[6] Since the EU ban, additional studies have been done which have suggested that there may also be links between exposure to DEHP and both obesity and type 2 diabetes in adults, as well as to allergies in children. And there is also suspicion that many wild species are also experiencing endocrine disrupting effects from phthalate plasticizers. These findings are not, however, completely conclusive, and as of this writing, US regulatory agencies are taking a cautious approach, with the Food and Drug Administration suggesting only that DEHP-containing medical products should be avoided in male infants.[7]

Bisphenol A

The chemical bisphenol A (BPA) was first synthesized in 1891 by the Russian chemist Aleksandr Dianin, although it was not manufactured commercially until the mid-1930s, when it began to be used to make epoxy resins. Even then, however, it was not produced in really large quantities until the 1950s when it started to be used to make the new plastic polycarbonate, a strong material that was synthesized by chemically joining together large numbers of BPA molecules. Polycarbonate was independently discovered in 1953 by two industrial chemists, one working at Bayer in Germany and the other at General Electric in the United States, and it soon proved to be one of the strongest, clearest, most shatter-resistant synthetic materials ever produced. The two companies both began marketing it in 1958 under the proprietary names Macrolon and Lexan. Since then, it has become ever more widely used in applications requiring high impact strength, optical clarity and ease of forming. Uses include auto headlight covers, safety glasses, fighter airplane cockpit canopies (it can be made bullet-resistant), hard plastic drinking bottles, CDs, DVDs and protective equipment such as police riot shields.[8]

In addition to being the main component of polycarbonate, BPA is also a key ingredient in epoxy resins, which like polycarbonates, have also become more widely used in recent years. Common applications for them include structural plastics (they are often used in the huge blades of wind turbines which generate renewable electricity), adhesives, paints and lacquers. Epoxy coatings are also widely used in food packaging, although this is not generally disclosed on product labels. When foods are canned in uncoated metal containers, the combination of water, heat and any acids they contain (think tomatoes) can work together to corrode the inside walls of the can, adding impurities to the food and changing its flavor. To prevent this, the inside of metal food cans are commonly coated with temperature resistant lacquers, which are largely based on epoxies. If these lacquers break down chemically, they can release some of the BPA that was used to make the epoxies in the first place, and under these conditions, small amounts of BPA may leach into the can and be consumed with the food.

Bisphenol A began to make headlines in the 1990s as the result of some studies suggesting that it had the potential to cause harm to children and adults as the result of its presence in many consumer products like hard plastic bottles and the liners of food cans. It also seemed to pose a threat to wild creatures since significant amounts were also ending up being released into the environment. Like DEHP, BPA has chemical similarities to estrogen and can interact with cellular estrogen receptors; both synthetic substances belong to a group of molecules sometimes known as *xenoestrogens*, or foreign estrogens. Laboratory experiments showed that high doses of BPA could alter normal sex hormone concentration and function in laboratory animals, and suspicions arose that it could also cause similar damage in people and wild vertebrates, especially aquatic creatures. In amphibians, BPA has been demonstrated to interfere with the development of tadpoles into frogs by disrupting thyroid hormone pathways, while in fish it increases the likelihood of undesirable cross-species mating (hybridization).[9] Other studies even suggest that it interferes with the capture of atmospheric nitrogen by bacteria living on the roots of plants of the pea family, a process that is important for the maintenance of fertility in both natural plant communities and agricultural systems.[10]

In humans, however, the evidence that BPA poses a significant threat is not clear-cut, with *hawks* and *doves* on either side of the debate and some government agencies in the middle. The Endocrine Society, an international group of endocrinology professionals including medical doctors, university researchers and others, believes that BPA is definitely "a significant concern to public health."[11] It can be detected in the urine of more than 90 percent of Americans, and some physicians, scientists and public health professionals think that it may also do harm to other body systems beyond the reproductive system, including the heart, brain and thyroid gland. There has even been speculation that exposure to BPA increases the risk of obesity. On the other side of the debate, detailed information marshalled by a chemical industry sponsored organization called the Polycarbonate/BPA Global Group states that "consumer products made with BPA are safe for their intended uses and pose no known risks to human health."[12]

Some countries have taken action to restrict the presence of BPA in certain consumer products. In 2008, the public health agency in Canada, for example, determined that BPA was a toxic substance and banned its use in baby bottles. This was followed by outright bans or health warnings in several European countries as well as in several states in the US, although health officials in a number of other countries took the position that based on the information known so far, there was not enough evidence to warrant a ban on BPA even in food containers for babies and children. In the US, for example, the National Institute of Environmental Health Sciences has expressed only cautious concern about BPA, while calling for further research.[13] In light of all the publicity, manufacturers of a range of consumer products—starting with baby bottles, adult water bottles and other food containers—voluntarily began replacing polycarbonate in their products, or else switched to other polycarbonate-like plastics that were not synthesized from BPA. In some countries, epoxy-based can liners were replaced by other kinds of plastics. Some manufacturers are also taking steps to replace BPA with a related chemical, bisphenol S, which is thought to be less potentially harmful.

So far, it is hard to draw any final conclusions about BPA and human health—it is possible, for example, that the amounts ingested from epoxy can linings and other food sources really are safe. Other sources of human exposure, however, have not been well studied, particularly its prevalence in thermally printed sales receipts which contain significant amounts of BPA that might be absorbed through the skin. For the rest of the living world, it clearly poses a greater threat, and a gradual phase-out seems to make sense to limit the future damage to a wide range of wild creatures from amphibians and fish to nitrogen-fixing bacteria and the many plants that depend on them.

Toxic Metals

Up to this point, we have examined two groups of pollutants which attack the body in fairly specific ways: organophosphate pesticides that block the normal functioning of nerve synapses; and hormonally-active chemicals used in plastics and plastic manufacturing which disrupt the normal balance of key hormones, interfering with reproduction and development and sometimes also increasing the likelihood of cancer. Other types of toxic substances, however, may have even more widespread harmful effects because the biochemical damage they do can affect a very broad range of cells, tissues and organs. Some of the best examples of this are provided by toxic metals like arsenic, mercury and lead. Before people began to refine metals from ore rocks, these three substances were relatively rare in ecosystems, and they never became part of normal vertebrate body chemistry. (Several other metals—iron, zinc, copper, manganese, chromium, molybdenum and cobalt—*are needed* by the body in trace amounts, but arsenic, mercury and lead are not.) The growth of technology since prehistoric times, however, led to ever greater releases of these three problematic metals in both pure and chemically combined forms. Metal refining was the first source, followed by the burning of coal. And then, from the eighteenth century on, industrialization and the steady expansion of chemical manufacturing led to dramatic increases in the use and release of these three toxic substances.

Arsenic

People have known about substances containing arsenic for thousands of years, although it was only identified as a distinct element in the Middle Ages. It is a widespread but fairly uncommon component of the earth's crust which exists in several different chemical forms. The earliest reports of its use both as a medicine and a poison date back roughly 4,000 years, and at least since the days of ancient Rome, different arsenic formulations have been particularly well known as poisons, a use they were especially suited for because they were tasteless and odorless. Since then they have featured prominently in struggles for aristocratic dominance and succession, in illegitimate acquisition of property, in suicides, and in the termination of unhappy marriages. In China, by around AD 900 arsenic compounds also saw some use as insecticides. The doctrine of the Renaissance physician Paracelsus that "the dose makes the poison" seems particularly true for arsenic. Since it is widely present in the environment, most people have had some exposure to it. Yet at low doses it has little if any harmful effect, while at high doses it is usually fatal. Some researchers have even speculated that it may in fact be a trace nutrient required in miniscule doses for good health, although no proof of this has yet been found.[14]

The modern history of arsenic compounds can be traced back to 1775 when the pioneering pharmacist/chemist Carl Wilhelm Scheele discovered copper arsenite, a bright green compound that others later used to make a dye known as Scheele's green. In the following decade, some physicians in Europe also began using weak preparations of potassium arsenite (Fowler's solution) to treat syphilis and leukemia. Around 1814 another arsenic-based green dye, copper acetate tri-arsenate, began to be prepared and eventually it came to supplant Scheele's green because it was thought to be less toxic. It was clearly still hazardous, however, and was given the name Paris green after strong formulations of it were used in that city as a rat poison. By 1867, Paris green had started to be used in the United States as an insecticide against the Colorado potato beetle. In the 1870s and 1880s it was also tried against the introduced gypsy moth that was devastating eastern North American oak forests, although it was found to also damage the trees themselves. By 1892, Paris green had begun to be replaced by another arsenic-based insecticide, lead arsenate, which was less toxic to plants, although for a while the two were often used in combination.[15] After that, until the development of DDT in 1939, lead arsenate was a standard insecticide worldwide, especially to control insects feeding on orchard fruit. It use resulted in numerous poisonings and deaths among agricultural workers and in residual lead and arsenic remaining on fruit sent to market. Where orchards were sprayed year after year, local groundwater supplies were also often contaminated.

Once industrial dye-making began in the middle of the nineteenth century, arsenic compounds became even more common and were used both as dyes themselves and as chemical reagents in the dye-making process. This sometimes led to dye workers being poisoned, along with ordinary citizens living near the factories, while many river ecosystems in Europe became devastated by toxic discharges. Throughout the nineteenth century and into the twentieth, preparations containing arsenic were widely available and commonly used in consumer products like wallpaper, as well as in inks and dyes, patent medicines and cosmetics. Looking back we can clearly see that their widespread use produced a steady trickle of poisonings and deaths—Scheele himself died in early middle age almost certainly from the toxic effects of arsenic and other hazardous substances that he had frequently experimented with.[16]

During the nineteenth century, people seem to have had a contradictory view of arsenic. It was well known that large doses caused acute illness and death with symptoms of headache, stomach pain, diarrhea and vomiting. Yet there seems to have been little suspicion that low-level exposure to arsenic-containing products could also be hazardous. At a time when spoiled, contaminated and adulterated food was all too common, food-poisoning like symptoms were not at all unusual, and arsenic's complex and often confusing effects made it harder for people to understand its risks. Until the 1830s, this made it even more attractive to poisoners, although the development of a sensitive chemical test for arsenic in 1836 by James Marsh made it much more likely that murderers who employed it would eventually be punished.

Chemically speaking, arsenic has similar properties to phosphorus, the element above it in the periodic table, and one that plays several important roles in cells. Except for a few kinds of bacteria, arsenic is not known to be an essential nutrient for living things, and it is normally absent from animals and plants, or else present only in minute quantities. Phosphorus, on the other hand, is critically important to all living cells because phosphate molecules are essential components of DNA and RNA, of vertebrate bones and teeth, and of cellular energy system molecules like ATP (adenosine tri-phosphate). Much of arsenic's toxicity arises from the body's mistaking it for phosphorus: it combines with enzymes which normally interact with phosphorus compounds to short-circuit some of the critical energy pathways inside cells. Exposure to arsenates and arsenic oxides cause a wide range of illnesses which can affect the nervous system, the liver, the heart and the kidneys; significant exposure can cause death from multiple organ failure, and it is also implicated as a cause of a range of different cancers.[17]

Up until the middle of the twentieth century, much of the work of public health personnel in industrialized countries focused on preventing infectious disease and malnutrition, with the main weapons being provision of clean drinking water, mass immunizations and addition of vitamin and mineral supplements to food. By the 1970s, however, as ecology and environmental toxicology became established as rigorous, scientific fields, and as the prevalence of infectious disease declined, public health initiatives focused increasingly on identifying threats from toxic substances and on controlling human and environmental exposure. After several decades of research, allowable exposure levels for numerous toxic substances began to be set by law, including levels of harmful metals like arsenic, mercury and lead. This, however, turned out to be an on-going project, since the enormous complexity of industrial society's chemical ecosystem, coupled with continued progress in medicine, biology and environmental health, meant that the hazards posed by toxic substances were continually being reassessed. As new information emerged, it was often the case that acceptable limits of exposure had to be revised downward. In the US, for example, mid-twentieth-century standards for allowable arsenic in drinking water were originally set at fifty parts per billion. As research continued into arsenic's harmful effects, however, by the turn of the twenty-first century the USEPA had decided to lower allowable arsenic levels in drinking water by 80 percent to 10 parts per billion. The Agency had in fact initially sought a level of only 5 parts per billion but came to accept, after public comment, that reducing arsenic to that level would be very costly for many water suppliers to achieve. In Bangladesh and a number of other countries where groundwater supplies have unusually high levels of arsenic—often over 50 parts per billion—widespread low-level arsenic poisoning currently affects tens of millions of people.[18]

In the early 1930s, at a time when lead arsenate was a standard farm pesticide, another innovative use was found for arsenic, this time as a wood preservative in combination with copper and chromium—the chromium was a fixative, the copper a fungicide and the arsenic an insecticide. Developed in India, copper chromium arsenate (CCA) soon began to be used to treat telephone poles in the United States and other countries, where the resulting material came to be known as *pressure-treated wood*. After World War II, it became very widely used in residential construction as wells as for docks constantly exposed to water. It was even used to build many playgrounds. It was remarkably resistant to decay and even termite attack under most circumstances, surviving intact for decades under conditions where most untreated lumber would have rotted. This gave it considerable economic value and brought rot-resistant wood down to a price range where millions of people could enjoy outdoor decks, wooden stairs and the like. It was even used in some conservation areas for boardwalks over wetlands to allow better access and education for the public.

Towards the end of the twentieth century, however, it was discovered that over time some of the arsenic was liable to leach out of the wood where it could be absorbed either by the skin of people who came into contact with it or by the surrounding soil. It was particularly prone to do this in acid soils. The result of this information was a ban in the United States on CCA treated wood for most residential uses beginning in 2003. Chemically treated wood was not abandoned as a product, but different copper compounds began to be used which did not contain arsenic. In addition to restrictions on treated wood and drinking water, human exposure to arsenic is regulated by standards that have been set for workplace air as well as for foods such as pork, eggs and poultry. Such food standards are needed because arsenic-containing compounds are frequently added to livestock feed to kill parasitic worms.[19]

Mercury

Mercury is the rarest of these three toxic metals in the earth's crust, although its compounds have been known for thousands of years because its most common mineral, mercuric sulfide, or cinnabar, has a striking red color. This made it sought after as a pigment, which came to be known as vermilion. It sometimes also occurs as pure metal, where its unique appearance as a shiny liquid makes it attractive as a curiosity. Its first limited uses seem to have included cosmetics and medicines, and by early historical times ground cinnabar had begun to be dissolved in lacquer to make red Chinese lacquerware. The Romans also knew of mercury and mined it near Almaden, Spain, although in Roman and later times, long-term work in the mercury mines was broadly understood to be a death sentence, and so convicts, forced laborers or slaves often made up the majority of the workforce. After the 1550s, the use of mercury exploded once it was discovered that mixing it with other chemicals and adding the mixture to low-grade silver ore allowed silver to be recovered. This created tremendous demand and led to large-scale mining in the small number of locations where large cinnabar or metallic mercury deposits occurred.

Metallic mercury itself is not as toxic as many of the compounds it forms. Of these, organic mercury compounds are generally the most dangerous, with methyl mercury being one of the worst. Even low levels of it, as from eating certain kinds of fish, is believed to cause neurological damage, particularly in developing fetuses; a number of studies have also found that it significantly reduces reproduction in fish, birds and mammals.[20] Mercury's toxicity began to be studied and understood in the first half of the twentieth century. Manufacturers of felt hats had long used mercuric nitrate to create a felted surface on animal skins that were used in hats, but by the 1940s, the hazards of this became clear, and the use of this substance in hat-making was banned.[21] Starting in the early nineteenth century, mercury also began to be used in dental fillings which were made up of roughly 50 percent metallic mercury and 50 percent silver and other metals. Some health authorities have argued that very small amounts of mercury commonly enter body tissues as a result of this practice, and a few countries have banned it since substitute filling materials have become available. But the mainstream of dentistry has long argued that such metallic *amalgam* fillings are perfectly safe due to the chemical and physical stability of the metal mixture.

Today, mercury release and human mercury exposure are controlled in most countries, although in some developing countries where it is still mined, significant poisoning of mine workers may still be occurring. Beyond the mining of mercury ore itself, the greatest releases into the environment come from the burning of coal, and from occasional large volcanic explosions which can eject significant quantities of mercury vapor into the atmosphere. The burning of coal and other fossil fuels is the largest man-made source, which is one reason that coal-burning is opposed by so many in the environmental community. Other significant releases into the atmosphere come from metal smelters and cement plants. Metallic mercury was also once used widely in thermometers and other measuring instruments, as well as in electrical devices, but most such uses have now been phased out due to health concerns, although tiny amounts are still employed in fluorescent light bulbs. Mercury's toxicity is well documented and is based on its interference with the normal functioning of the trace element selenium. Selenium has a long history in living things and is present in cells as a component of numerous essential enzymes which prevent internal oxidative damage and also carry out other functions. Mercury severely disrupts many of these essential selenium enzymes and is particularly damaging to the brain, nervous system and kidneys.[22]

In the past, most mercury poisoning occurred in miners, smelter workers, instrument makers and manufacturers of felt hats, all of whom were directly exposed to mercury or its minerals. Today, most of those occupations have disappeared or been modified to protect workers' health. The mercury released by volcanic eruptions is limited by the relatively small number of continually active volcanoes. Mercury from industrial pollution, or from large-scale burning of coal, on the other hand, is more problematic, especially when it ends up in aquatic habitats. When this happens, mercury undergoes a process of concentration in food webs due to the inability of animals to excrete it. In most places, the element exists in very low concentrations in water, but becomes a bit more concentrated in algae which also convert it into a more toxic organic form. When the algae are, in turn, eaten by tiny water animals, most of the mercury ingested is retained, concentrating it further. This process is repeated with small fish eating the tiny water animals, medium-sized fish eating the small fish, and finally, large, long-lived fish and marine mammals eating the medium-sized fish. People, unfortunately, tend to have a strong preference for eating species which are aquatic predators at the top of the food chain like swordfish and tuna from the ocean and bass and pike from fresh water. It is these fish which contain the most organic mercury, and which, if eaten, pose the greatest health risk, especially to pregnant women and their babies.

Workplace mercury exposure is now well-regulated in most developed countries, and laws, regulations and voluntary compliance by manufacturers have led to dramatic reduction in its use in batteries, measuring instruments and other categories of products. Large amounts are still, however, dispersed by coal-burning, metal smelting, cement making and other industries, and the significant costs involved in reducing these sources has made further reduction of toxic emissions a slow and difficult process.

Lead

As we saw at the beginning of the previous chapter, lead was the first metal people actually smelted from rocks, and it must have been a fascinating material given its shiny, soft, moldable qualities. It is not a particularly common element in the earth's crust, although it is more abundant than either arsenic or mercury. The first things people made out of lead were beads, and then small statues. By 5,000 years ago lead had also begun to be incorporated into pottery glazes, and not long after it began to find uses in construction. By classical Greek times, the first *bullets*—small football-shaped projectiles designed to be thrown from slings—were being made from it, and not long after, in ancient Rome, lead began to be utilized on a very large scale, so much so that widespread lead poisoning may have occurred. Roman industries carried out extensive mining and refining of metals, and they obtained lead not only directly from lead ore but also as a by-product of silver refining. They used it in dozens of different ways as water pipes, plumbing fixtures, tanks and cisterns, roofing and construction materials, cooking vessels, food additives, cosmetics and so on. It was widely employed wherever easy shaping and/or heavy weight, corrosion resistance and water-tightness were desirable properties. Following Rome's collapse—perhaps due in part to extensive lead poisoning of its elite—lead use decreased, although it did not end, and in medieval times it continued to be used for pottery, roofing, waterproofing, and the like, while also taking on new roles as the supporting frameworks for stained glass windows, and as a starting material for alchemists' experiments.[23]

After the invention of firearms in twelfth-century China, lead found another important use. Guns needed something to shoot, and the heavy, soft, shapeable qualities of lead made it seem ideal for bullets and *shot*. By the fifteenth century, metallic lead had also begun to be used in printing in the form of movable type, and brightly colored lead salts such as white lead (lead carbonate) and chrome yellow (lead chromate) were also starting to be used as pigments in paints and cosmetics. In the eighteenth century, the beginnings of industrialization saw many new applications for this heavy metal such as the lead-lined tanks that John Roebuck used to manufacture sulfuric acid on an industrial scale. Other new uses continued to be found throughout the nineteenth and the first half of the twentieth century. In the nineteenth century, lead paint pigments began to be more widely used as production expanded to meet the needs of a growing middle class, making paints brighter, more durable and more resistant to temperature changes. After 1813, the development of metal food cans also created demand for a low-temperature solder to seal the cans, and lead began to be used for that purpose, although by the early twentieth century, the dangers of lead poisoning from canned food had led to the replacement of lead solder by other metals. Advancing industrialization throughout the nineteenth century also dramatically lowered the cost of different kinds of metal pipes, and soon iron, brass and copper lines were being used to provide all manner of indoor plumbing, including drinking water pipes. And even though improved manufacturing increased the precision of pipes and fittings, numerous sealed joints were still needed, with lead coming into wide use as solder and packing to ensure water-tightness.

In 1859, the rechargeable lead-acid battery was invented in France to provide a new source of portable electric power. Its design was improved in the 1880s, and it soon proved crucial to the newly developing automotive industry because it provided a practical, low-cost way of repeatedly producing and storing enough electricity to start cars and to run their ignition and headlights. Over the course of the twentieth century, lead-acid batteries came to be produced on a truly amazing scale. By the first decade of the twenty-first century, for example, more than 800 million automobiles were in service around the world, almost every one equipped with a lead-acid battery, with most batteries containing between 15 and 20 pounds of lead.[24] Total battery lead alone therefore amounted to roughly fourteen billion pounds. Current annual world lead production exceeds six billion pounds, and each year roughly nine or ten billion pounds of new and recycled lead are put into new batteries and other products, an amount roughly fifty times as great as the Romans are estimated to have used in a year.[25]

Beyond batteries, automobile and truck tires needed to be balanced on their wheels in order to allow high speed operation, and this has usually been accomplished by attaching lead weights of varying sizes to the outside of wheel rims. The majority of these weights remained attached to the wheels and could be recycled when new tires (and new wheel weights) were installed. Inevitably, however, many thousands of lead weights were lost into the environment every year from the hundreds of millions of vehicles with balanced tires. In a few countries, efforts are under way to switch to zinc and other less toxic wheel balancing weights, but in the first decade of the twenty-first century, an estimated fifty million pounds of lead was still being used each year to make new wheel balancing weights.[26]

Most of the lead that is present in automotive batteries is currently reprocessed and reused, both to protect the environment and to save money, and this has dramatically reduced the potential pollution from it. But in some countries people working with lead, or living close to places where lead has been processed or used, are still being poisoned during the second decade of the twenty-first century. In China, for example, which produces and consumes more lead than any other nation, and leads the world in lead-acid battery production, the organization Human Rights Watch issued a 75-page report in 2011 describing the poisoning of thousands of Chinese children by emissions from lead smelters, lead-acid battery plants and workshops making "tin-foil".[27] Their report claimed to be based on information gathered in Hunan, Shaanxi, Yunnan and Henan Provinces, and concluded that China was facing a widespread lead poisoning epidemic. They also reported massive efforts to hide this widespread poisoning both from the international community and from many of the victims themselves, who it claimed were being denied access to lead testing for both adults and children.

Paints were another leading use of lead throughout the nineteenth and most of the twentieth century, with painters and paint factory workers at particularly high risk for harm. White lead and lead chromate did improve the durability of paint films, yet over time paints containing them still tended to flake and crumble, particularly around windows and doors. When old paint was scraped off to allow repainting, even more lead was released. Outside, it tended to build up in the soil around houses; inside, it slowly accumulated as dust and small flakes within the living space. The Romans had known the sweetness that lead sometimes imparted to their food and wine, and that same flavor in paint chips made them attractive to toddlers who routinely put all sorts of non-food items in their mouths. Once placed in the mouth, the amount of lead that toddlers can absorb from paint chips rises dramatically. In the mid-nineteenth century, this became an important, slow-acting source of lead intoxication in people, a problem that still persists today. Public health authorities started to recognize the dangers of lead paints for children beginning in the 1890s, leading to its restriction in Queensland, Australia in 1897. This was followed by bans in several European countries in 1909 (France, Belgium, Austria). Most other countries, however, did not immediately follow suit. The United States, for example, only began to phase out lead-based house paint in 1971, and surveys carried out in the early twenty-first century found that lead-based paints were still being manufactured and sold in numerous countries, especially in South Asia. Even as recently as 2007, millions of children's toys that had been made and exported from China were recalled in the United States and other countries because they either contained lead, or were painted with lead paint.[28]

In the nineteenth and twentieth centuries, so much lead was mined that in some cases locations with large deposits often became intensely polluted. One such area is in North America near the joint borders of the states of Oklahoma, Missouri and Kansas. Here, in a 40-square-mile area known as the Tar Creek Superfund Site, much of it within a Native American reservation, extensive lead and zinc mining was long carried out, and decades of digging left enormous piles of heavy-metal laden mine wastes dotting the landscape. Contamination of ground and surface waters along with dust blowing from the waste piles have resulted in varying degrees of lead poisoning in local people, especially in children. On one section of the site, around the communities of Picher and Cardin, Oklahoma, the land had been so badly contaminated and so honeycombed with tunnels that the US Government evacuated both towns at great expense, buying up all the houses and relocating the residents. There are hopes that most of the huge piles of tailing containing lead, zinc and cadmium will eventually be incorporated safely into paving materials used in road construction. But it is unlikely that large areas of the Tar Creek Site will ever be suitable for normal human habitation or that the ghost towns of Picher and Cardin will ever again see human residents.[29]

Lead-acid batteries played an important part in the widespread introduction of trucks and automobiles beginning at the end of the nineteenth century. It was not until the 1920s, however, that lead in another form came to be used in large quantities in automobiles, this time as a fuel additive designed to smooth engine operation and allow more powerful motors. The story begins with the work of Thomas Midgely who was one of America's most celebrated industrial chemists during the first half of the twentieth century. Midgely was born in 1889 in Pennsylvania, and studied mechanical engineering at Cornell University. After graduation in 1911, he worked first at the National Cash Register Company and then in his father's automobile tire business. In 1916, he joined the Dayton Engineering Laboratories Company which was run by the American automotive pioneer Charles F. Kettering, a co-inventor of electric starters for cars—an invention made workable by lead-acid batteries. Internal combustion engines had originally been designed as stationary power plants able to take advantage of the new petroleum-based fuels that were coming into the market. One thing that limited their workability, however, was a tendency to rough combustion, which at the extreme could lead to premature component wear or even to severe engine damage. Often this was caused either by incomplete burning of fuel followed by small, ill-timed after-explosions, or by premature combustion of the fuel–air mixture as it was being compressed in the cylinder.

At Dayton Engineering Kettering asked Midgely to search for new gasoline additives that could improve the smoothness of combustion, a difficult problem which led to several years of research and experiments. At that time, it was already known that addition of ethyl alcohol to gasoline reduced rough combustion. But problems with that additive, including increases in metal corrosion, had limited its usefulness. In the middle of this research period however, World War I diverted much of the Dayton lab's expertise to military work, and Midgely's efforts came to focus on industrial chemistry, specifically on the search for high octane gasoline suitable for ultra-high compression airplane engines which were always at risk of having their fuel ignite prematurely due to compression heating. By the time the high octane aviation fuel project was successfully completed, utilizing a mixture of additives of cyclohexane and benzene (a substance which was later found to be carcinogenic), World War I had ended and Midgely was able to return to his research on anti-knocking additives for lower compression automobile engines. For two or three years, beginning around 1919, he and his co-workers systematically tried a large number of different chemical additives until they found one that was relatively inexpensive and performed exceptionally well in their tests—tetraethyl lead.[30]

Midgely's team made the new additive by combining a chlorinated hydrocarbon with a sodium-lead compound, and it did a good job of eliminating rough combustion. It was not the only suitable material. Various other substances could do it as well, but they had to be added to gasoline in larger quantities. It took only relatively small amounts of tetraethyl lead to do it, and after being patented it became a very profitable product. It did, however, have two significant problems— it physically fouled engines with lead oxide; and it was highly poisonous—Midgely himself developed a serious case of lead poisoning from his exposure to it and had to undertake a long convalescence in 1922–1923. To avoid engine fouling, tetraethyl lead came to be routinely mixed with dichloro- and dibromoethane, compounds that reacted with the lead inside the engine as the gasoline burned to produce lead chloride and lead bromide. These were then released into the environment along with the exhaust and formed the actual toxic emissions. (Thomas Midgley is also remembered for having been part of the team at General Motors that first developed chlorofluorocarbon refrigerants such as *Freon*. They did this while searching for a substitute for the hazardous refrigerants like ammonia, propane and sulfur dioxide, that were being used at the time. They were successful in so far as chlorofluorocarbons were not flammable, grossly toxic or explosive. But from an environmental point of view they also failed spectacularly because the new refrigerants eventually turned out to damage earth's ozone layer which protected life on the planet from harmful ultraviolet rays.[31])

Around 1920, Kettering's company merged with the recently created General Motors Corporation (GM), and Dayton Engineering became GM's research division. Once Midgely and his team had developed tetraethyl lead, General Motors began preparing for a major effort at manufacturing and marketing the product and to do this they shared samples with a variety of industrial and academic scientists and engineers. Immediately, there was controversy because several leading academic experts were convinced that its lead content made tetraethyl lead highly toxic and that it should not be mass-produced for use as a fuel additive. The Surgeon General's office in the United States was one of those expressing deep reservations, but at the time it had no legal authority to intervene. Some testing was undertaken by the United States Bureau of Mines, but the Bureau admitted in internal documents that the industry had forced special conditions on the testing which placed it in a conflict of interest position since the Bureau depended on industry cooperation to carry out its core mission. In the meantime, General Motors had made a decision to manufacture tetraethyl lead in cooperation first with the DuPont chemical company and then with Standard Oil.

In 1923, GM's own production started in Dayton, Ohio, while Du Pont began manufacturing the additive in Deepwater, New Jersey. Soon, workers making the product in both Dayton and Deepwater began to die of lead poisoning—one in 1923, five in 1924 and four in 1925, leading GM to shut down the Dayton operation. In mid-1924, a third production line opened at Standard Oil's Bayway, New Jersey oil refinery, but after only a few months five workers were dead and dozens more had been hospitalized, leading to a temporary shut-down there. At that point, the corporate leadership of the three companies became very worried—in that era, worker illness after years of service was not considered a big problem, but numbers of almost immediate deaths and hospitalizations from a new production line might not only make it impossible to successfully carry out manufacturing, but could also generate torrents of bad publicity and perhaps even state-level regulatory action.

In the end, however, the huge potential profits to be made seem to have overcome all these concerns, and the companies launched a sales and marketing campaign for *ethyl*—the word *lead* was not mentioned—in which they did everything they could to convince the public that their product was safe, even hiding significant evidence indicating that it might not be. Politicians went along, and public health bodies of the day lacked the authority to directly regulate tetraethyl lead's manufacture or use. Following the tragic factory deaths and injuries, improvement in worker safety seems to have been led by DuPont, the partner with the most experience in the mass production of hazardous chemicals. Their representatives are reported to have been shocked at the tremendous employee exposure involved in the hands-on Standard Oil production line, and they notably improved safety by adopting a closed process which dramatically reduced direct worker contact. Large-scale production began in 1925. By the end of the 1930s, gasoline made with tetraethyl lead had achieved an almost complete dominance of America's gasoline market, and was also well on its way to doing the same throughout the rest of the world.[32]

No records were kept of the exact tonnage of lead dispersed into the environment from gasoline additive between the 1920s and the end of the twentieth century, by which time most gasoline no longer contained lead. It is, however, clear that vast amounts of the toxic substance were emitted. Statistics available around the time of the phase-out in the United States make it possible to at least gauge the rough magnitude of the releases: in 1979 alone, 177 million kilograms—more than 177,000 tons—of lead are believed to have been emitted from motor vehicles just in the United States, and that was after a 1971 phase-out order had already reduced leaded gas use by roughly one third. Using this figure as a benchmark, it seems fairly certain that at the minimum, millions of tons of lead were emitted worldwide from gasoline additives in the last three quarters of the twentieth century.[33]

Tetraethyl lead in gasoline was the single largest creator of global lead pollution, but many others sources were also important, including one that directly affected wildlife all over the world: lead released in hunting, shooting and fishing. Lead bullets, shot and fishing sinkers were not new inventions in the nineteenth century, but the rise of more and more powerful and easier-to-use rifles and shotguns, especially the invention of self-contained cartridges, coupled with tremendous growth of population and disposable income, led to a dramatic increase in the use of firearms and fishing tackle. Again, no accurate long-term records were kept about the amount of lead used to produce ammunition and fishing tackle, but even at the end of the nineteenth century, enough lead was being blasted around for observers to note deaths among water birds from swallowing spent shot. In fact, More than 500 scientific studies published since 1898 have documented that, worldwide, 134 species of wildlife are negatively affected by lead ammunition.[34]

Lead from bullets, bird shot and fishing sinkers is persistent in the environment. The only way it is effectively removed from ecosystems is by deep burial under sediment and debris, a process which occurs very slowly in most habitats. Ever since Roman times, and much more rapidly since the nineteenth century, the amount of lead present in forests, fields, marshes and water bodies has continually been increasing. As with gasoline additive and paint lead, no precise overall statistics are available of the amount of lead in all the bullets, shot and fishing weights used since the nineteenth century, but the magnitude of such sport releases can be estimated from surveys carried out in recent years. Roughly 60,000 metric tons of lead was released every year by hunting and target shooting in the United States during the first decade of the twenty-first century, and that figure does not take into account lost lead fishing weights.[35]

By the end of the twentieth century, it had become widely accepted that lead poisoning had reduced populations of ducks and geese, along with those of loons and other water birds. In response, in 1991, lead shot was finally banned in waterfowl hunting in the United States. Lead has also had negative impacts on the populations of dozens of other species of wildlife, but proposed bans on lead bullets, shot and fishing sinkers have been met by vociferous opposition from sportsmen worried about the higher cost of non-lead ammunition and the potential for greater wear and tear on firearms. Modern warfare has also been a major source of lead pollution, along with the release of numerous other toxic substances, and it seems certain that at least hundreds of millions, and probably billions, of lead bullets were dispersed by the world's military personnel during the nineteenth and twentieth centuries.

Mechanisms of Lead Toxicity

The idea that lead was toxic has been around for more than 2,000 years. It appears in the writings of Nicander, a Greek botanist who lived in the second century BC, and of Vitruvius, a Roman architect of the first century BC. Vitruvius, for example, wrote that it was much healthier to drink water delivered through clay pipes than through lead ones. Roman civilization nonetheless made very extensive use of lead, not just in water pipes, but also in pots and jugs used to cook food and process wine. Some scholars in fact believe that chronic, widespread lead poisoning caused increased incidence of madness and infertility within the Roman elite and was a significant cause of the collapse of their civilization.[36]

After the fall of Rome, the amount of lead mined and used in Europe dropped considerably, and most poisoning was concentrated in workers involved in mining or smelting the metal, in stained-glass window making, and in a handful of other professions. By the 1600s, physicians had begun to document lead poisoning among miners and metal workers, and in the 1700s colic and other gastrointestinal symptoms were often caused by drinking alcoholic beverages—wine or rum, for example—which had come into contact with lead during their manufacture. In the nineteenth and early twentieth centuries, as we have seen, widespread industrialization led to ever more uses for lead, exposing more people, and polluting more places. At the same time, new science-based fields of public health and occupational medicine were starting to develop and to focus on the significant health dangers presented by lead. Researchers began to make systematic descriptions of the effects of both short-term, high-level lead exposure—which was often fatal—and of lower-level, chronic exposure. Yet a real understanding of lead's widespread toxic impacts on both human populations and on wild creatures was still not possible, because limited insight into the deeper working of cells and molecules meant that no one could really explain how lead actually damaged the body in so many different ways.

By the second half of the twentieth century, however, this had begun to change as cellular and molecular biology came of age. By then, physicians and public health professionals were able to document more than one hundred different kinds of harm that lead could do to unborn babies, infants, children and adults, including damage to the nervous, digestive, reproductive, circulatory and excretory systems. In children, for example, it could cause mental retardation along with dozens of other medical problems. Lead also tended to accumulate in the body, notably replacing enough of the calcium in bones for its presence to be detected as dark shadows on X-rays. Biochemists noted that even though it was a much heavier element than calcium with a very different location in the Periodic Table of the elements, lead had several key chemical properties in common with that lighter, biologically important element. This fact turned out to be one of the keys to understanding the widespread damage it could cause in the body.[37]

For many years, it has been widely understood that calcium was important for human (and animal) health, if only because it is a major constituent of bones and teeth. But like several other mineral nutrients such as sodium and potassium, calcium also turned out to have other important functions in humans and animals beyond that of a building block for hard tissues. In the last decades of the twentieth century, as the enormously complex inner workings of cells began to be systematically explored, calcium was discovered to play key roles in a surprisingly wide range of different chemical pathways, especially those involving signals sent internally within cells. As we will see below in our description of cancer cells, healthy cells are constantly receiving biochemical messages from neighboring cells as well as from cells in other parts of the body. Most such signals come in the form of hormones and other molecules which arrive at a cell's outer surface or membrane where they stimulate receptor structures that connect to the inside of cells. Once a particular surface receptor has been triggered, it releases one or more chemical signals inside the cell in order to influence the range of products the cell will make over the next interval of time. After long and intensive research, it was discovered that many of these internal signaling substances require calcium in one form or another in order to function properly. In fact, nearly 200 such calcium-dependent internal signaling molecules have already been discovered, and they are known to play vital roles in a variety of important cellular processes.[38]

In contrast to calcium, lead normally plays no role in body functioning, and it is difficult for the body to eliminate it once it has been absorbed. Part of the threat it poses is based on the fact that lead resembles calcium closely enough for the body to confuse the two, although not closely enough to function in place of calcium in critical biochemical pathways. As a result, if lead is present, the production and working of calcium-dependent internal signaling molecules is disrupted in cells.

This is not, however, the only way that lead causes harm. It also has a tremendous affinity for sulfhydryl groups, chemical units which contain a hydrogen and a sulfur atom. Sulfhydryls are present in one of the amino acids and play a critical role in allowing cellular enzymes to fold up into fully functional shapes. Some of these enzymes normally counter the effects of *free radicals*, cellular chemicals that have become unusually reactive. Some free radicals are always produced by metabolism, but cellular enzymes usually neutralize them quickly. If lead is present, however, sulfhydryls bind strongly to the metal, making it impossible for the body to make enough functional protective enzymes. This allows free radicals to hang around longer and do much more harm to cells. These are the two main mechanisms that allow lead to cause damage to a wide range of tissues in people and animals.[39]

Over the second half of the twentieth century, understanding of lead's harmful effects led public health officials in many countries to define safe upper limits for lead in the human body, particularly in children. Several times over those decades, however, the growing body of medical knowledge has required revisions downward in the amount of body lead deemed safe. Eventually, by the end of the twentieth century, it was finally realized that *any and all* lead in the body can cause harm, and that the only truly safe level of internal lead is *zero*.[40]

Carcinogens—Substances That Cause Cancer

Given the complexity of the structures and biochemical pathways inside our bodies and the many thousands of ways that different molecules interact within them, it is not hard to understand that some toxic substances like organophosphates cause damage in narrowly specific ways, while others, like lead, are able to disrupt a much wider range of bodily processes. What is harder to make sense of is the way that some chemicals can harm living tissues not just by disrupting *current* biochemical pathways, but by damaging the underlying biochemical recipes or *genetic codes* which regulate and control cells over time. This can happen in several ways including direct damage to DNA such as breaks in the strands or insertion of molecules derived from pollution that interfere with DNA's normal functioning. It can also result from changes in the living *text* that DNA provides for cells through the insertion of an incorrect DNA subunit, the deletion of a correct one, or both.[41] Some DNA changes cause only limited damage. But others, especially if a number of them accumulate in the same cell, can lead to serious abnormalities and even to uncontrolled growth of defective cells. Some individuals are also born with a number of undesirable DNA changes already in place, changes that can interact with new damage to make things worse. When this happens, *cancer* is often the result, a term that is used to describe a broad category of serious and often fatal cellular diseases.

Our current understanding suggests that several different factors can stimulate the development of cancer by causing damaging changes in DNA: ionizing and ultraviolet radiation (see Chapter 8); chemical carcinogens; viral infections; and chance biochemical events. These causes are not mutually exclusive, however, and in many cases the badly mutated cells that give rise to tumors or leukemia may actually have been damaged by a combination of toxic chemicals, radiation, and accidental errors in gene copying. Living cells, however, do not just passively accumulate damage. Billions of years of cellular evolution have produced elaborate mechanisms inside cells for identifying DNA damage and repairing it, mechanisms that are continually at work in people and in other organisms. These are partic-ularly effective if damage has been done to just one strand of double-stranded DNA because the intact strand allows the restoration of the correct information. If DNA should be damaged beyond repair, mechanisms related to repair can also often shut down the dysfunctional cell or even have it commit suicide (apoptosis), so it will not develop into a tumor. In addition, animal immune systems, which evolved over time to defend against infection, include mechanisms for identifying and destroying damaged cells, including cancer cells. If these mechanisms are working well, early cancer may be nipped in the bud without any symptoms being experienced. But if mutations or other changes weaken our immune responses, serious disease has a greater chance of occurring.[42] In general, genetic damage to cells tends to accumulate over time, and so older people and animals are more likely to develop tumors than younger ones—even some ancient dinosaurs suffered from tumors, as X-rays of fossil bones have clearly shown.[43]

It is hard to really make sense of what is going on in cancer except in light of the history of life. For hundreds of millions of years, all living things were single cells. But then in several different groups of unicellular creatures, a number of lineages evolved *group living*. Some that did this remained simply colonies where there was only minimal coordination and signaling among members. Others, however, went down a path of deeper cooperation in which large numbers of cells came to function cooperatively as a single organism. This allowed their groupings to grow larger and to develop highly specialized cell functions that dramatically increased overall colony efficiency. Biologists believe that multicellular creatures arose several times in the history of life, although so far the fossil record offers only limited clues to the details of these events. What is known is that multi-celled creatures with different types of tissues were present by 550 million years ago, and that eventually they evolved into the different kinds of animals, plants and fungi living on earth today.[44]

The tremendous success of multicellular animals, plants and fungi attests to the survival value of this approach. Yet for the individual cells involved there was also a price to pay. The ancient journey of living things on planet earth began with single cells that were able to absorb chemical nutrients, synthesize critical substances, grow in size, and then reproduce by dividing in two. Today, billions of years later, uncountable numbers of bacteria, archaea and other single-celled organisms still live this way. For such creatures, a single cell is truly a whole organism, and each one is largely on its own as it attempts to survive and reproduce; neighboring cells may in fact pose a serious threat because they compete for the same nutrients and the same living space. In general, single-celled organisms feed and reproduce as much as they can, and the daughter cells they produce are mostly just like themselves. Most commonly, they restrict their feeding and reproduction only when conditions in the environment are unfavorable to them.

Cells living in multi-celled organisms, on the other hand, face a very different set of imperatives. Each one is normally able to survive only as long as the larger organisms to which it belongs remain alive, and this stark fact ends up changing some of the basic ground rules of cellular life. As *team players*, cells in larger organisms take on a particular shape and metabolism during the period when the organism is developing—in animal embryos, a given cell may turn into a nerve cell, a muscle cell, a skin cell, a liver or kidney cell, or one of dozens of other types. Each body cell of a multicellular organism starts off with the same set of genes. But as development goes on, the majority of the genes in any given cell are deactivated in order to allow specialization appropriate to different kinds of body tissue. The small set of genes that are functional produce only the specific cellular anatomy and chemistry that goes along with being one particular type of cell. And even once *differentiated* in this way, the lives of such cells are still very different from those of single-celled creatures. Being part of larger multi-celled creatures, they no longer carry out unlimited cellular reproduction once the organisms to which they belong are fully grown. Instead, they continually communicate biochemically with surrounding cells and regulate their activities so as to keep their section of living tissue in a stable state. And in order to do this, they often engage in two kinds of behavior that would not make much sense in a single-celled organism: limiting their own cell duplication or reproduction, and committing cellular suicide.

In most animal tissues, when cells sense that they are crowded, they begin to divide more slowly. In extreme cases, some will even collapse and die in a neat, controlled way, a process known to scientists as programmed cell death or *apoptosis*. This happens routinely during the development of embryos if cells are in the way of important structures that are taking shape—the lines of skin cells that join the fingers into a webbed flipper in human embryos, for example, all die during normal development so that babies are born with five separate fingers and toes on each limb. And some of the billions of nerve cells initially present in the developing brain die off as fully functioning nervous pathways develop near them. In addition, cells sometimes commit suicide if they suffer certain kinds of biochemical damage. So critical and well-designed are these genetically controlled breakdown mechanisms, that they can actually benefit the larger creature they belong to. Unlike cells damaged by physical injury or by bacterial or viral infections, cells that commit suicide for the good of the organism do so in an orderly, clean, step-by-step way that even allows some of the useful substances they contain to be recycled into other cells. In recent decades, biologists have discovered several different mechanisms for programmed cell death—one common process involves the release inside the cell of powerful enzymes which are able to chemically disassemble essential cell proteins. Apoptosis, however, remains a very active area of research.[45]

Every part of the ancient mechanisms for maintaining cellular stability in tissues depends on the presence of intact genetic instructions. Mutations can interfere with cells correctly sensing the density of surrounding cells, with lowering their rate of cell division when crowded, with adhering properly to one another, and with their ability to undergo apoptosis when appropriate. In addition, mutated cells are also often unable to carry out the normal functioning of their tissue type—mutated liver cells, for example, may no longer be capable of adjusting the amount and type of fats and carbohydrates dissolved in blood or of filtering out harmful substances. If enough mutations accumulate to serious disrupt normal functions and apoptosis or removal by the immune system does not occur, the cell in question can turn into a full-fledged cancer cell which is often capable of dividing and dividing until it kills the organism in which it is living.

The complex details of how specific kinds of cancers develop in response to different genetic damage from toxic substances or ionizing radiation is very much an on-going area of research. Cells have many genes and there is a wide range of tumors and toxic chemicals—cigarette smoke alone, for example, contains thirty-three different identified carcinogens.[46] There is also more than one type of ionizing radiation; and the body's mechanisms for DNA repair and programmed cell death can be disabled to varying degrees. In a few cases, however, the nature of the damage has clearly been identified. Significant exposure to vinyl chloride, for example, a key substance in the making of PVC plastic, is well documented to cause liver cancer, especially tumors of a type called angiosarcomas. It manages to do this in one of two ways, either by combining chemically with cellular DNA so as to alter its workings, or by physically breaking apart chromosomes, the DNA packaging structures that are critical to normal cell functioning.[47]

There is also one last level of complexity, which is that some toxic or cancer-causing substances only begin to cause damage after they have been chemically altered inside organisms. Animals often attempt to deal with unfamiliar molecules by decomposing the original substances in organs such as the kidneys or liver. When this happens, a range of breakdown products is produced, and in some cases, it is these so-called *metabolites*—they can be thought of as biochemical debris—that end up causing the most damage, including the induction of tumors. Benzene, for example, is a mass-produced, toxic industrial chemical that formerly was present in significant quantities in gasoline, and even was marketed directly to consumers as a cleaning product. It is absorbed into the body by breathing in fumes from the air, and once taken in, it ends up being processed by the liver. For decades it has been known to attack bone marrow, causing leukemia (a cancer of blood cells), and anemia, a shortage of red blood cells. Until the end of the twentieth century, however, the mechanism by which it caused cancer was not at all clear. Finally, however, advanced biochemical research began to show that some of the initial breakdown products it generated in the liver, particularly benzene oxide, were among the specific toxic compounds that caused genetic damage to bone marrow cells.[48]

Regulation, Policy and Public Awareness

This chapter and the previous one have focused on selected chemical technologies, on the scientific advances that made them possible, and on their impacts on the environment and on human health. But people, of course, were also a key part of these innovations. The thoughts, beliefs and economic motives of the businessmen, financiers, political leaders, scientists and engineers whose work created chemical technologies played just as important a role in their development as did the practical techniques—the financing, the physical infrastructure and the scientific theory—which allowed chemical production to increase. Many of the nineteenth and early twentieth century leaders of chemical research and manufacturing saw themselves as rational people pursuing technological advancement for the benefit of humankind (while perhaps also making a good living for themselves). Yet the deeper roots of their thinking, and of their desire to control and manipulate the world around them, may well have originated not in a nineteenth-century factory, laboratory or management office, but rather farther back in time, in the thousands of years of agricultural civilization which preceded industrialization.

Before settling down as farmers, our foraging ancestors seem to have had only limited dreams of controlling and manipulating the world. They frequently set fires in the landscape to favor vegetation that supported their preferred prey animals, but in general they seem to have viewed themselves as part of the natural ecosystems through which they ranged. But once their descendants started to repeatedly cultivate crops in the same location, their basic attitudes towards nature likely started to shift. Hunter–gatherers had relied on the survival of natural systems to provide their future security, and the last thing they sought was to completely eliminate whole ecosystems. But agriculturalists' way of life was fundamentally different, and they essentially became competitors with natural systems for land, water, sunlight and nutrients. Foragers tended to move around over a wide range, seeking food and resources as they seasonally became available. Permanent farmers, in contrast, stayed put, especially in places like river valleys where water and fertile soil was generally available most of the year. And once settled, they began to view many other creatures that occupied the same land—plants and animals alike—as threats. Often, the majority of local animals that could not be hunted for food were reclassified as *varmints* or *pests*, while local plants that were not of direct use to people or livestock came to be considered as *weeds* or *brush*. In this way, native creatures were often turned into adversaries, things to struggle against and to strive to overcome. This mindset eventually came to dominate much of agricultural civilization, and it continued right through a century or two of industrialization and up to today, in spite of the fact that technology has fundamentally changed the economic base of society while environmental science has provided a far more accurate understanding of humankind's true relationship to nature.

Before the scientific revolution, insightful individuals realized that they could not understand detailed aspects of nature well enough to predict or control what was likely to happen in the future. But to some, modern science and mathematics seemed to have changed all that. Kepler and Newton had made it possible to precisely predict the paths and locations of the planets and moons. In the following centuries, similar observational and mathematical tools allowed one emerging field of science and technology after another to gradually develop the capacity to predict and control some of the things they studied. The result, by the nineteenth century, was both a sense of hopefulness for a brighter future for a human species which had long been plagued by poverty, hunger, disease and death, and also a dream of limitless power and the ability to master nature through the use of science and technology. The first goal was frequently referred to as the doctrine of *progress*; the control of nature was generally thought to be necessary if human advancement was actually to continue.[49]

Full scientific understanding, of course, turned out to be a much larger undertaking than even the early astronomers and physicists could have imagined. Biology in particular, including medicine, public health and ecology, turned out to be enormously complex subjects requiring many decades before even the outlines of what there was to know had clearly emerged. And sadly, in many cases like those of chlorofluorocarbon refrigerants, human knowledge of how to create and use powerful technologies developed well in advance of our understanding of the far-reaching and damaging impacts that those technologies could have on living things.

In the nineteenth- and early twentieth-century chemical industry, the ideas of progress and the conquest of nature joined with personal ambition—and sometimes greed—to drive the leaders of chemical technology forward to great accomplishments, while also making them over-confident and relatively indifferent to the harm they might be doing to human health and the rest of nature. One result was that they often tolerated conditions in their factories that sickened or even killed their own workers, while at the same time releasing toxic discharges and emissions that polluted areas near and far. In this, though, it is fair to say that the chemical industry in those days was little different from other workplaces, and that exposure to harmful substances in chemical plants was probably not any more hazardous in general than the dangers workers faced in other settings such as factories, railroads, mines and mills.

In the state of Massachusetts in the United States, for example, the building of the nearly five-mile long Hoosac railroad tunnel between 1855 and 1875 cost the lives of 193 workers, while between 1870 and 1883, 27 workers died in New York City building the Brooklyn Bridge; and in Monongah, West Virginia, on December 6, 1907, more than 350 coal miners died in a single mining accident.[50]

By the beginning of the twentieth century, an ever wider array of new synthetic substances was being manufactured, along with increased amounts of older materials like lead. The engineers and chemists in charge of these operations were highly trained in physical science and math, but generally had little knowledge of medicine or ecology, which in any case were still in their infancy. Lacking such knowledge, and in the absence of either government regulation or public awareness, little or no thought was given to the subtle, far-flung and long-term effects of the many compounds that were being discovered and manufactured. In the nineteenth-century alkali industry, it had been quite obvious that the hydrogen chloride coming out of the soda plants was corrosive and harmful. But in the late nineteenth- and early twentieth-century petrochemical industries, it took a long time before it became equally clear that exposure to even moderate quantities of barely detectable benzene could cause workers to develop fatal cancers like leukemia.

By the late 1920s, advances in biological and medical understanding of illness and disease, and the creation of new fields of industrial health and hygiene, had started to highlight the dangers of many new chemicals. But the Great Depression of the 1930s and World War II in the 1940s cast whole societies into desperate struggles for survival, making environmental health and safety seem like luxuries that no one could really afford. For the vanquished in World War II, and even for some of the victors like Russia and Britain, the immediate postwar period was one of continuing struggle and privation. In the United States, however, the boom in chemical production and research and development that had characterized the war years continued after 1945, giving the country an important lead in new products and materials. US chemical manufacturing and exports soared, and worldwide output also increased as revitalized industries in Europe and Japan steadily increased their own production.

By the 1960s, the world was awash in manufactured chemicals. Their widespread use without meaningful testing or understanding of their complex effects had begun to cause serious degradation to ecosystems and wildlife populations, with some groups of living things such as birds of prey and amphibians (whose moist, absorbent skin and aquatic habitats made them particularly vulnerable) experiencing heavy losses. Around the same time, however, the rapidly developing fields of ecology and environmental health were beginning to accumulate evidence that some widely used manufactured chemicals were causing significant damage to human health and to natural ecosystems, both locally and globally. These findings helped set the stage for the rise of the environmental movement in the 1960s and 1970s. In some cases, regulations and limitations governing the use of individual harmful substances started to be imposed, although usually not before considerable damage had already been done to people and/or ecosystems. In other cases, implementation of restrictions on chemical use was blocked by politicians in response to lobbying by powerful economic interests representing the petroleum, chemical or agricultural industries. In many poorer countries, a low level of education and a lack of effective governance often led to continued use of highly damaging substances long after they were restricted in richer countries.

Even in the face of rising environmental awareness, the development, manufacture and use of new synthetic and toxic chemicals continued at a rapid pace throughout the second half of the twentieth century. Many harmful substances continued to see widespread use, although a small number did eventually come to be regulated or banned as being particularly toxic. (In the United States, for example, the persistent chlorinated hydrocarbon insecticide heptachlor can still be used in underground electrical transformers to control imported fire ants, although it is banned for all other uses.) International efforts in this direction culminated in the Stockholm Convention of 1995, which banned some of the most damaging substances including a number of chlorinated pesticides. Besides being toxic and/or cancer-causing in varying degrees, most of the compounds proscribed in Stockholm were persistent and tended to become increasingly concentrated in animals higher up in food chains. Additional toxic synthetics that contained chlorine were also eventually banned, especially compounds of the dioxin-furan group which included the infamous PCBs.[51] These steps represented significant progress, although they still failed to address the broader issue of human society and the natural environment being bathed in a flood of manufactured chemicals whose overall impacts remain very poorly understood.

Even in the early twenty-first century, thousands of manufactured substances, including carcinogens, neurotoxins, endocrine disruptors and toxic metals are still fully legal to use in many settings in many countries. These continue to pose significant threats to human health and to natural systems, often in insidious ways.

First, their effects are often felt far from their initial point of use or release, both in space and in time, turning many in to regional, if not global pollutants. Second, it is prohibitively expensive to submit every single synthetic industrial chemical to adequate testing, and in any case, standard assays for human and environmental toxicity are limited to testing individual, pure substances, rather than the mixtures of compounds normally encountered in the real world. Third, the natural environment and human beings everywhere have already experienced the release of thousands of different manufactured chemicals over more than a century, many of them in enormous quantity. Some evidence has begun to suggest that quite low levels of different combinations of them can be harmful, even though separate exposure to small amounts of each individual substance might be quite safe.

One final fact that seems to emerge clearly from the history of manufactured chemicals is that in most societies the level of knowledge of human biology, ecology, and science in general are critically low and woefully inadequate. Effective education in these areas might not only impart greater understanding of specific vulnerabilities and risks, but also might lead to a changed mindset in which nature is viewed as an intricate, ancient and essential thing—an ally of humanity—rather than as an enemy to be conquered using every weapon at our disposal.

Notes

1   M. B. Colovic et al., 2013, Acetylcholinesterase inhibitors: pharmacology and toxicology, *Current Neuropharmacology* 11(3): 315–335, www.ncbi.nlm.nih.gov/pmc/articles/PMC3648782; Wikipedia, Acetylcholine, https://en.wikipedia.org/wiki/Acetylcholine.

2   National Pesticide Information Center, Malathion: technical fact sheet, http://npic.orst.edu/factsheets/archive/malatech.html.

3   Wikipedia, Parathion, https://en.wikipedia.org/wiki/Parathion.

4   European Environment Agency, Neonicotinoid pesticides are a huge risk—so ban is welcome, says EEA, www.eea.europa.eu/highlights/neonicotinoid-pesticides-are-a-huge.

5   E. Diamanti-Kandarakis et al., 2009, Endocrine-disrupting chemicals: an Endocrine Society scientific statement, *Endocrine Reviews* 30(4): 293–342.

6   Wikipedia, Bis(2-ethylhexyl) phthalate, https://en.wikipedia.org/wiki/Bis%282-ethylhexyl%29_phthalate.

7   US Food and Drug Administration, DEHP in plastic medical devices, www.fda.gov/MedicalDevices/ResourcesforYou/Consumers/ChoosingaMedicalDevice/ucm142643.htm.

8   Wikipedia, Polycarbonate, https://en.wikipedia.org/wiki/Polycarbonate.

9   R. A. Heimeier et al., 2009, The xenoestrogen bisphenol A inhibits postembryonic vertebrate development by antagonizing gene regulation by thyroid hormone. *Endocrinology* 150(6): 2964, http://endo.endojournals.org/content/150/6/2964.full; J. L. Ward, and M. J. Blum, 2012, Exposure to an environmental estrogen breaks down sexual isolation between native and invasive species, *Evolutionary Applications* 5(8): 901–912, http://onlinelibrary.wiley.com/doi/10.1111/j.1752-4571.2012.00283.x/abstract;jsessionid=AD2835A9AD807A4EFE76313A177FA515.f01t02.

10 J. E. Fox et al., 2007, Pesticides reduce symbiotic efficiency of nitrogen-fixing rhizobia and host plants, *PNAS* 104(24): 10,282–10,287, www.pnas.org/content/104/24/10282.full.pdf.

11 E. Diamanti-Kandarakis et al., 2009, Endocrine-disrupting chemicals: an Endocrine Society scientific statement, *Endocrine Reviews* 30(4): 293–342.

12 Bisphenol-A.org, About bisphenol A, www.bisphenol-a.org/about/index.html; American Chemistry Council, About BPA: weight of scientific evidence supports the safety of BPA, https://plastics.americanchemistry.com/Product-Groups-and-Stats/PolycarbonateBPA-Global-Group/About-BPA-Weight-of-Scientific-Evidence-Supports-the-Safety-of-BPA.pdf.

13 US National Institute of Environmental Health Sciences, Bisphenol A (BPA), www.niehs.nih.gov/health/topics/agents/sya-bpa.

14 J. M. Dolliffe, 1993, A history of the use of arsenicals in man, *Journal of the Royal Society of Medicine* 86: 297, www.ncbi.nlm.nih.gov/pmc/articles/PMC1294007/pdf/jrsocmed00098-0055.pdf; M. F. Hughes et al., 2011, Arsenic exposure and toxicology: a historical perspective, *Toxicological Sciences* 123(2): 305–332, http://toxsci.oxford-journals.org/content/123/2/305.full.

15 F. J. Peryea, Washington State University, 1998, Historical use of lead arsenate insecticides, resulting soil contamination and implications for soil remediation, http://soils.tfrec.wsu.edu/leadhistory.htm.

16 Wikipedia, Paris Green, https://en.wikipedia.org/wiki/Paris_green.

17 World Health Organization, Arsenic, www.who.int/mediacentre/factsheets/fs372/en.

18 U. K. Chowdury et al., 2000, Groundwater arsenic contamination in Bangladesh and West Bengal, India. *Environ Health Perspectives* 108(5): 393–397, www.ncbi.nlm.nih.gov/pmc/articles/PMC1638054.

19 Wikipedia, Arsenic, https://en.wikipedia.org/wiki/Arsenic.

20 Wikipedia, Methyl mercury, https://en.wikipedia.org/wiki/Methylmercury.

21 Wikipedia, Mercury (element), https://en.wikipedia.org/wiki/Mercury_%28element%29.

22 US Agency for Toxic Substances and Disease Registry, ToxFAQs for mercury, www.atsdr.cdc.gov/toxfaqs/TF.asp?id=113&tid=24; Wikipedia, Mercury poisoning, https://en.wikipedia.org/wiki/Mercury_poisoning.

23 Wikipedia, Lead, https://en.wikipedia.org/wiki/Lead.

24 Answers.com, How much lead is in a car battery?, www.answers.com/Q/How_much_lead_is_in_a_car_battery.

25 International Lead Association, Lead production, http://ila-lead.org/lead-facts/lead-production; Wikipedia, Lead, https://en.wikipedia.org/wiki/Lead; Wikipedia, Lead-acid battery, https://en.wikipedia.org/wiki/Lead%E2%80%93acid_battery.

26 USEPA, 2008, News release—hazardous waste, http://yosemite.epa.gov/opa/admpress.nsf/3ee0a48cce87f7ca85257359003f533d/85f6da001abb40a885257506008 2293f!OpenDocument&Highlight=2,lead.

27 Human Rights Watch, 2011, My children have been poisoned: a public health crisis in four Chinese provinces, www.hrw.org/en/reports/2011/06/15/my-children-have-been-poisoned-0.

28 Wikipedia, Lead poisoning, https://en.wikipedia.org/wiki/Lead_poisoning; Reuters, 2009, Lead-laden paint still widely sold around the world, www.reuters.com/article/2009/08/25/us-lead-paint-idUSTRE57O64G20090825.

29 US Agency for Toxic Substance and Disease Registry, 2016, Tar Creek Superfund site—Ottawa County, Oklahoma, www.atsdr.cdc.gov/sites/tarcreek; Wikipedia, Tar Creek Superfund site, https://en.wikipedia.org/wiki/Tar_Creek_Superfund_site.

30 National Academy of Sciences of the United States of America, Biographical memoir of Thomas Midgley, Jr. 1889–1944 by Charles F. Kettering presented to the academy at the annual meeting, 1947, www.nasonline.org/publications/biographical-memoirs/memoir-pdfs/midgley-thomas.pdf.

31 Ibid.

32 B. Kovarik, 1994 (revised 1999), Charles F. Kettering and the 1921 discovery of tetraethyl lead in the context of technological alternatives, originally presented to the Society of Automotive Engineers Fuels & Lubricants Conference, Baltmore, MD, www.environmentalhistory.org/billkovarik/about-bk/research/henry-ford-charles-kettering-and-the-fuel-of-the-future.

33 D. J. Hoffman et al., 2002, *Handbook of Ecotoxicology*, second edition, CRC Press, p. 377.

34 US National Park Service, Lead Bullet Risks for Wildlife & Humans, www.nps.gov/pinn/naturescience/leadinfo.htm; D. J. Pain et al., US National Park Service, 2009, A global update of lead poisoning in terrestrial birds from ammunition sources, www.nps.gov/pinn/learn/nature/upload/0108%20Pain.pdf; R. T. Watson et al. (eds.), 2008, *Ingestion of Lead from Spent Ammunition: Implications for Wildlife and Humans*, conference proceedings, The Peregrine Fund, www.peregrinefund.org/subsites/conference-lead/2008PbConf_Proceedings.htm.

35 Science Daily, 2004, Do lead bullets continue to be a hazard after they land?, www.sciencedaily.com/releases/2004/11/041104005801.htm.

36 H. Delile et al., 2014, Lead in ancient Rome's city waters, *PNAS* 111(18): 6,594–6,599, www.pnas.org/content/111/18/6594.full.

37 Wikipedia, Lead poisoning, https://en.wikipedia.org/wiki/Lead_poisoning.

38 R. H. Garrett and C. M. Grisham, 2006, *Biochemistry*, third edition, Brooks/Cole, p. 1060.

39 US Agency for Toxic Substance and Disease Registry, Toxicological profile for lead, www.atsdr.cdc.gov/toxprofiles/tp.asp?id=96&tid=22; K. Nemsadze et al., 2009, Mechanisms of lead-induced poisoning, *Georgian Medical News* 172–173: 92–96, www.ncbi.nlm.nih.gov/pubmed/19644200.

40 Wikipedia, Lead poisoning, https://en.wikipedia.org/wiki/Lead_poisoning.

41 Wikipedia, DNA repair, https://en.wikipedia.org/wiki/DNA_repair.

42 Ibid.

43 L. C. Natarajan, 2007, Bone cancer rates in dinosaurs compared with modern vertebrates, *Transactions of the Kansas Academy of Science* 110: 155–158, http://arxiv.org/ftp/arxiv/papers/0704/0704.1912.pdf.

44 K. J. Niklas, 2014, The evolutionary-developmental origins of multicellularity, *American Journal of Botany* 101(1): 6–25.

45 A. G. Renehan, 2001, What is apoptosis, and why is it important? *BMJ* 322(7301): 1,536–1,538, www.ncbi.nlm.nih.gov/pmc/articles/PMC1120576.

46 Wikipedia, List of cigarette smoke carcinogens, https://en.wikipedia.org/wiki/List_of_cigarette_smoke_carcinogens.

47 US Agency for Toxic Substances and Disease Registry, Toxicological profile for vinyl chloride, www.atsdr.cdc.gov/toxprofiles/tp.asp?id=282&tid=51.

48 R. Snyder, and C. C. Hedli, 1996, An overview of benzene metabolism, *Environmental Health Perspectives* 104(Suppl. 6): 1,165–1,171, www.ncbi.nlm.nih.gov/pmc/articles/PMC1469747; S. M. Rapaport et al., 2010, Human benzene metabolism following occupational and environmental exposures, *Chemico-Biological Interactions* 184: 189–195, http://superfund.berkeley.edu/pdf/329.pdf.

49 M. Meek Lange, 2011, Progress, *Stanford Encyclopedia of Philosophy*, http://plato.stanford.edu/entries/progress.

50 Wikipedia, Hoosac tunnel, https://en.wikipedia.org/wiki/Hoosac_Tunnel; West Virginia Division of Culture and History, Monongah mine disaster, www.wvculture.org/HISTORY/disasters/monongah03.html; Wikipedia, Monongah mining disaster, https://en.wikipedia.org/wiki/Monongah_mining_disaster.

51 Stockholm Convention, http://chm.pops.int/Home/tabid/2121/Default.aspx; Wikipedia, Stockholm Convention on Persistent Organic Pollutants, https://en.wikipedia.org/wiki/Stockholm_Convention_on_Persistent_Organic_Pollutants.

6

POWER TECHNOLOGIES

Background

Our hunter–gatherer ancestors were fit, active people who went around on foot. They frequently moved their camps to follow wild game or to take advantage of seasonally available plant food, and when they did, they generally had to carry everything they owned, sometimes for long distances and over difficult terrain. Even if they remained in one location, foraging for plants involved a lot of walking, carrying, cutting, and digging, while hunting and fishing required walking, running, throwing things, cutting up game and carrying prey back to camp. These were important kinds of work that our pre-agricultural ancestors performed, along with the making of clothing, tools, and personal possessions, caring for one another, preparing food and engaging in ritual and recreational activities. In most cases, they got a great deal of exercise. The most important aspect of their lives, however, in which overwhelming physical force might have been of greatest use—the capture and killing of prey animals—was usually accomplished not through raw power, but through ingenious technologies and learned hunting skills. These included cooperative hunting strategies, sharp points and cutting blades that could injure or kill animals without first physically over-powering them, and devices like traps, snares and weirs that could hold prey captive until hunters returned.

As we saw in Chapter 3, the transition to farming from hunting and gathering marked a revolutionary change in the way human beings lived. It settled people on land cleared of natural vegetation and committed them to the endless cycles of work required for cultivation: soil preparation, planting, weeding, harvesting and processing their crops. Farmed land generally tends to decline in fertility over time as the result of erosion and the repeated removal of nutrients by crops. This made never-ending cultivation even more important, and the continual heavy work of loosening the soil, weeding and harvesting meant that physical labor was often in short supply. Eventually, some of this need was met by draft animals, although these required pastures, water and food. And millennia were to pass before early farmers discovered effective ways of harnessing oxen, horses and water buffalo to the plough. And even after draft animals had been integrated into farming operations, agriculture still required so much hard human labor that it was all too common for much of the work to be done by human beings condemned to back-breaking lives of serfdom or even outright slavery. For thousands of years, endless, hard, physical labor completely dominated farming, and even today it remains the norm in some places and for certain crops.

During this long period, numerous ingenious hand tools and larger mechanical devices were invented to help with many tasks, although they did not fundamentally alter the basic routines of continual, physical work. Nor was it possible in most places for people to ever return to hunting and gathering. Land clearing meant that most of the wild animals and plants that hunter–gatherers had relied on for food were no longer there. And even where this was not the case, greatly expanded human populations could no longer be supported solely by harvesting wild foods. Against this background, any additional sources of power which could assist agriculture were extremely welcome.

Flowing water was probably the first force used to supplement muscle power, although for a long time its use was limited to delivering irrigation water to the fields by directly diverting rivers and streams that were running downhill from surrounding uplands. Besides flow irrigation, running water eventually found other applications. Water-driven wheels fitted with shelves or buckets began to be used to lift water vertically to irrigate higher fields, reducing the human and animal labor involved (although where streams to run the wheels were not available, they were often powered by gangs of laborers or draft animals). By the third century BC, water-powered mills were also being used to produce baking flour by passing whole grain between massive, rotating grindstones. These were widely employed in ancient Rome, often driven by the rush of water flowing downhill through constructed aqueducts. Besides grinding grain, water-powered mills were also used to saw wood and to run hammers for iron-making.[1]

By 5,000 years ago, wind energy had also begun to be captured by sailboats transporting cargoes, especially those moving upstream against the current on the river Nile. Sails had no immediate effect on food production, but they did provide a boost to food distribution since boats offered the only efficient way to move bulky or heavy cargoes long distances. These capabilities reduced the chances of outright starvation by allowing shortages in one place to be made up from surpluses in another. Not long after, in the Old World, ocean-going sailboats came to dominate the long-distance transport of heavy goods. By the ninth century AD, in the Near East, people had applied the principles of water-driven mills to breezy areas where windmills were built and used to raise water or to grind grain.

Farming, of course, was not the only important activity where agricultural societies felt the limitations of human (and animal) strength and endurance. Building was another labor-intensive activity. Wherever wood remained available for construction, felling trees, dragging them from the woods, hewing them into timbers or sawing them into boards all involved tremendous amounts of hard physical work for people and draft animals alike. And in drier lands, or where forests had long been leveled for crops and pastures, building with stone, brick or rammed earth also required enormous amounts of labor.

Steam

Looking back, it is easy to see that the scarcity of physical energy was a major limitation to the advance of civilization. What is harder to understand is how difficult it was to actually develop additional forms of power. After thousands of years of agricultural civilization, a truly important breakthrough in this area came only around 1700 with the development of the first practical steam engine. And even that revolutionary power source did not spring forth full blown from a single person's imagination. Rather, it emerged from decades of experimenting, theorizing and tinkering by a number of early inventors, scientists and mathematicians, all of whom were trying to understand two phenomena: the behavior of water moving through pipes and pumps, and the dynamics of steam inside containers. These inquiries, in turn, were part of broader scientific investigations being carried out on gases and water pressure. For the scientists themselves, the greatest interest lay in the fundamental questions involved. But in the spirit of the eighteenth century, practical technologies and useful new devices were also widely sought, and for this reason, scientists, inventors and other experimenters were constantly striving to design not only better pumps, but also entirely new kinds of devices powered by air, water, steam or vacuum. It had been known for a long time that enclosed steam could generate pressure; what had not been at all clear was how this energy might be harnessed to do useful work.

Pumping Engines

One of the earliest advances came in 1679 when the French scientist and inventor Denis Papin[2] demonstrated a device in London designed to steam bones at a high enough temperature to allow them to later be ground into bonemeal. His invention, which resembled a large pressure cooker, was fitted with a kind of safety valve, and as Papin operated his device, he observed that this valve moved forcefully upward as the pressure in the pot increased. This made a strong impression on his inventive mind, and by 1690 he had gone on to build a model steam engine containing a movable piston within a cylinder. Papin was a leading researcher in this period on devices powered by steam, water and air, and by 1704 he had even developed a primitive working steam engine that he used to run a paddle-driven boat. Despite his brilliance, however, he never managed to perfect any of his devices and ended up dying in poverty, although his work proved to be a crucial stimulus to other researchers and inventors, most notably the English blacksmith and iron merchant Thomas Newcomen (1664–1729).

Newcomen was from southwest England, a region where tin and other minerals had been extracted since the Bronze Age, and for years he had worked as an ironmonger, selling metal and metal tools. He was quite familiar with the local mining industry in part because he frequently visited local mines in order to sell tools to the miners.[3] Over time, mining activities inevitably led to the depletion of minerals near the surface. When this happened, tunnels had to be dug ever deeper into the earth until they reached far below the water table, effectively turning mine shafts into drainage channels. Water seeped into them continuously, and miners often had to work long shifts in partially flooded tunnels. To remedy this, primitive pumps and water wheels had been employed since Roman times to raise the water to the surface, although such devices were not all that effective and required enormous amounts of labor to operate.

With these problems in mind, Newcomen set himself the task of developing a better way to remove water from mines, and by 1712, he had designed a steam-powered mine pump based loosely on a design by Denis Papin. His machine used the weight of an underground pumping apparatus to pull down a large, overhead, rocking wooden beam which, in turn, raised a large vertical piston within a cylinder. The cylinder was then filled with steam at atmospheric pressure, after which cold water was sprayed inside to condense the steam in order to create a partial vacuum. This had the effect of forcefully sucking the piston back down, generating the engine's power stroke. In modern terms, Newcomen's device was incredibly inefficient: it burned tons of coal to pump modest amounts of water. But it *worked*. It was able to remove more water from greater depths than anything that had ever existed before, and as a result it quickly became a commercial success. In some mines, it replaced large teams of horses—sometimes 100 or more—that had been used to drive water wheels of a design that had not changed much since Roman times.[4]

Newcomen's steam engine was massive. It included a large stone pier in the middle to support the giant overhead beam, and a substantial brick firebox where fuel was burned beneath a boiler. Unlike the familiar gasoline and diesel engines of today, it was an *external combustion* engine—it burned its fuel not inside the cylinder in which the piston moved, but in a chamber outside it. Given this design, it was possible to use a wide range of combustible materials as fuel—wood, peat or even straw were used at different times, in different places. But most Newcomen pumping engines, and indeed most steam engines used throughout the eighteenth and nineteenth centuries, were fired by coal. Despite its great weight and relatively high cost, the Newcomen engine was the first truly successful machine of its type, and by the time its inventor died in 1729, he and his partners had constructed at least 75 of them at different mines around Britain. One other advantage the Newcomen engine offered was that it was located up on the surface rather than down in the mine shaft. This meant that the heat, noise and fumes it produced were vented straight into the open air, causing intense, local air pollution, but not directly poisoning the air inside the mine. In the decades following Newcomen's death in 1729, hundreds more were built, and by the early 1770s, the engineer–physicist John Smeaton had succeeded in at least doubling their efficiency by improving its design and using more precise manufacturing techniques and better seals to produce tighter-fitting pistons and cylinders.[5]

Improving the Piston Steam Engine

The success of the Newcomen engine was a critical step in launching the industrial revolution. Yet even with Smeaton's improvements, the mechanical power it produced represented just a tiny fraction of the total heat energy released by the fire in its boiler. In addition, its up-and-down motion and enormous size and weight limited its applications almost exclusively to pumping water out of mines. Its impact on the development of technology, however, was enormous. Decades of successful operation of Newcomen engines demonstrated clearly that steam could reliably be used to produce mechanical force, and this, in turn, led some of the most inventive minds of the second half of the eighteenth century to seek innovative ways of harnessing this new source of power. One of these was the Scotsman James Watt (1736–1819) who was a contemporary of important figures in the Scottish Enlightenment including the geologist James Hutton (1726–1797), the philosopher David Hume (1711–1776) and the economist Adam Smith (1723–1790). In Watt's time, Scotland was a place where science, progress and invention were all very much in the air.

As a youth Watt had trained as a maker of *mathematical instruments*, a term used in those days to describe scientific and other precision devices. After mastering his trade in London, he was hired in 1756 by the College of Glasgow to clean and repair astronomical instruments. One of his relatives was a professor there, and the noted chemist Joseph Black was also on the faculty. In 1763 or 1764 Watt was asked to repair a small, working model of a Newcomen type steam engine. He completely rebuilt it, but also discovered that its basic design was so inefficient that it was hard to keep running. Working in a research-oriented environment, Watt soon started to do quantitative experiments on the expansion of steam under different conditions. At times, he consulted Professor Black, seeking help, for example, in understanding the apparent *disappearance* of some of the energy in his experiments. As luck would have it, Black himself had only recently discovered the odd fact that as liquid water boiled to make less highly organized steam, some of the energy that drove this change seemed to disappear—the steam itself was not as hot as it should have been given the total heat produced by the fire, minus the losses to the air and to the experimental apparatus. The exact nature of this *latent heat* was not well understood until the twentieth century, but Black was able to advise Watt that the steam's measured temperature did not fully reflect all the energy it actually contained, a fact which laid the groundwork for some of Watt's own improvements in steam engine design.[6]

Watt then began an intensive study of steam engines, always with an eye to improving them. One insight that he gained was that the high fuel consumption of the Newcomen engine could at least partly be explained by the repeated dousing of the power cylinder with cold water in order to condense the steam.

Cooling the cylinder in this way meant that it had to be completely re-heated before the next power stroke was possible. This wasted a lot of heat and slowed down the machine. By 1765, Watt had come up with an ingenious solution to this problem in the form of an additional condensing chamber connected by a valve to the power cylinder. Once the steam had expanded and advanced the piston, it could then be led out into the second chamber to be condensed, generating a partial vacuum that sucked the piston back and provided some of the engine's power. The additional chamber allowed the power cylinder to remain continuously hot so that the next power stroke could occur sooner, and it ensured that less inlet steam was needed to move the piston because its heat was not lost in re-heating the cylinder.

Watt continued to work on steam engines from the 1760s through the 1780s, incorporating a series of improvements made by himself and his collaborators. Using technology initially developed for the making of military cannons, for example, he found that it was possible to bore far more accurate power cylinders, while improvements to metal machining allowed the outer surfaces of the pistons to be made more perfectly round. Combining these two advances made it possible to create a much tighter fit between the piston and the cylinder, reducing the leakage of steam and allowing an increase in the engine's power and efficiency. By 1775 Watt had produced an innovative *single-acting pumping engine* that was about four times as efficient as the best of the Newcomen designs. In other words, it could do the same amount of work using only one fourth as much coal.[7]

In the late 1760s Watt had also begun communicating with a leading British metal manufacturer and industrialist named Matthew Boulton (1728–1809), and by 1775 he had relocated to Birmingham, England to become Boulton's business partner. In their collaboration, the Scotsman contributed his talents in engineering and innovation, while Boulton offered business experience, production expertise and working capital. In the following years, as the company formed by the two men became the world's leading supplier of steam engines, they also incorporated a number of further advances into their designs. These included a more efficient double-acting piston and a sun and planetary gear transmission that was developed by William Murdoch, one of their leading engineers. The last made it possible for steam engines to drive rotating machinery. They also developed a boiler that worked at slightly above atmospheric pressure and a centrifugal speed governor (similar to a design previously used in windmills) which both reduced the chance of explosions and allowed the engines to operate with fewer constant adjustments by an operator.[8]

Generating rotary motion was critical, because it allowed steam engines to find nearly limitless applications in manufacturing and transportation. In the past, wind and water-powered mills had produced rotation which mostly was used to grind grain. But the ability to produce it on demand, wherever a steam engine could be located, proved to be a critical spur to industrialization. In the long run, of course, Watt's rigorous, science-based method of working proved to be just as important as the specific technical advances he made. Watt was a pioneer of modern mechanical engineering, a profession founded on scientific principles, systematic trials and mathematical analysis.

Applications of Steam Power

Textiles

In Chapter 4 we saw that the small number of factories existing in Europe in the early 1700s focused mostly on producing things like pottery, soap, metal goods, bricks, glass, and simple chemicals like alkali. In addition, large, specialized workshops in cities and towns by the sea built ships, although these were still crafted one at a time almost entirely by hand. At that time, except for the grinding of grain and the sawing of boards, which were usually carried out in water or wind-powered mills, almost all the other products that people produced—food, cloth, clothing, candles, furniture, books, jewelry and the like—were made on farms or in small workshops, using hand tools and simple muscle-powered machines. Enormous amounts of work were required to make many of these goods, and woven cloth was an essential product that was extremely labor-intensive to make.

As is the case with so many other things that we buy today, few of us ever give much thought to how our clothing is made, or where the cloth to make it comes from. Yet throughout both history and pre-history, people have invested enormous amounts of time, energy and creativity in efforts to discover better ways to transform raw fibers like wool or cotton into finished cloth and clothing. For natural fibers, the first major step is to straighten, align and clean the raw material. With wool, for example, this is done by washing and *carding*—a process which is named for a type of thistle, whose dried seed heads were initially used for this purpose. Next, the fibers had to be *spun*, in order to twist and link together the relatively short individual units into endless chains. If twisted tightly into thin strands, they made *thread*; if twisted less tightly into coarser strands, they made *yarn*. Most natural fibers are off-white in color (some wool is an exception), and if dark or brightly colored thread or yarn was desired, dyeing was also necessary. After that, the yarn or thread had to be wound onto spools or cones so that it could be stored or transported. Then it had to be unwound without tangling in order to be used for weaving, sewing or knitting. And even at that point, the thread, yarn and cloth that had been produced were still only the raw materials for making finished clothing and linens.

Until about 1700, most textiles were produced by craftspeople working in *cottage industries* in which family members labored cooperatively to produce basic textiles using simple, hand-powered tools such as drop spindles, spinning wheels and weaving frames. Under these conditions, it took enormous amounts of work to produce yarn, thread, cloth, clothing, carpets and knitwear. One result was that such products were relatively expensive, even though they were not always of the highest quality. After 1700, however, this all started to change as inventors in Great

Britain began to develop water-powered machinery capable of performing some of the many steps needed to turn raw sheep's wool or agricultural flax or cotton into yarn, thread and cloth. Many inventions and innovations followed, and by the middle of the nineteenth century, almost the entire textile-making process had been mechanized with muscle power being replaced first by water power and later by steam.

The first important step came in 1733 when the invention of the *flying shuttle* significantly increased the speed at which cloth could be woven. One impact of this, though, was that the yarn and thread that were the raw materials for making cloth were soon in short supply, since they were still being produced largely by hand. In response, inventors figured out how to speed up their manufacture, developing first a multi-spool device known as the *spinning jenny* (1764), then a water-driven spinning frame called a *water frame* (1768), and finally the *spinning mule* (1779), which combined features of both the water frame and the spinning jenny. Together, these sped up fiber processing so much that even large numbers of hand weavers equipped with flying shuttles were unable to keep up with the supply of yarn and thread that could now be produced. The next advance came in 1785 when Edmund Cartwright patented a power loom, although it was to take several additional decades before it became an efficient, fully workable device.[9]

As one mechanical innovation after another was developed in eighteenth-century Britain, the making of yarn, thread and cloth became more centralized in factories. As the new textile machines increased in size, complexity and number, there was also an increased demand for mechanical power to run them. Most of the earliest mills had been driven by a single water wheel, although the wheel, the canal and the necessary water gates were all expensive to build and required continual maintenance. Efficient factory production, however, generally required multiple machines in order to achieve a high output, so that the power shaft coming off the water wheel was often connected to a straight, overhead *line shaft* running the length of the factory. At intervals along it, long leather belts attached to pulleys led down to individual machines, allowing a single, large rotating shaft to power all of the factory's machinery. As the British textile industry grew, it eventually encountered a shortage of suitable streams with enough flow to reliably power a factory. This restricted the number of mills that could be built and spurred the search for new sources of power that were not dependent on unevenly distributed water resources.

The Newcomen engine was one obvious alternative to flowing water, but the fact that it did not generate rotary motion made it poorly suited to running the kinds of line shafts which powered most textile mills. By the mid-1770s, when Boulton and Watt made available their significantly improved steam engine, there was already considerable unmet demand for sources of power capable of running factory machinery. By the 1780s, when they began to sell engines specifically designed to provide rotary power, their new technology made it possible to build mills nearly anywhere that coal and fresh water were available, and this led to a significant increase in textile production. In America, for example, Newcomen engines had first begun appearing in the 1750s. But by the early 1800s, once the improvements to the steam engine made by Watt and others were incorporated, the number of installed units increased dramatically.

Transportation

Steamboats

Factories were not the only beneficiaries of improved steam engines. Denis Papin had built an experimental steam-powered boat at the beginning of the eighteenth century, and by the 1780s inventors started trying to adapt the Watt engine to power a ship. Even though Boulton and Watt's power plants were still quite heavy, they were much lighter than Newcomen's design, and the buoyancy of ships meant that their weight could be accommodated. In 1802, the first steamboat powerful enough to tow barges was built in Scotland, and in 1807 the American Robert Fulton used a Watt engine to power his North River Steamboat which provided a successful passenger service over the 120-mile route from New York City to Albany on the Hudson River.[10]

In those early days, nearly all steamships and boats were propelled by large, circular paddle-wheels placed on their sides. They were a bit like water-driven mills running in reverse—instead of having the water move the wheels, the wheels moved the water, propelling the boat forward. Steam-powered boats advanced steadily in the early decades of the nineteenth century, with their greatest initial use being on inland waters like rivers and large lakes where wind sailing was difficult.[11] By the 1840s, powerful steamships equipped with paddle wheels were also offering regularly schedule crossings of the Atlantic between Europe and North America, and from about 1845 on, ships powered by the much simpler and more efficient screw propeller came into service between North America and Europe. Soon, steam began to crowd out sail in long distance transportation and trade; on most routes it allowed faster average speeds, more direct routes and less dependence on the vagaries of wind and tide.[12]

Railways

Despite their reliance on highly variable winds, eighteenth-century sailing ships still represented very advanced and sophisticated technology; given favorable breezes, they were quite fast. On the land, by contrast, travel had not really improved since Roman times, and in many areas overland journeys were actually more difficult than they had been in the ancient world. In the middle of the eighteenth century, for example, wagon or coach travel on un-paved roads, even in relatively flat terrain, usually moved at a speed of just 4 miles per hour—about the pace of a brisk walk. At that speed, people could cover no more than about 40 miles over the course of a bone-jangling ten-hour day. By the middle of the nineteenth century, in contrast, trains managed speeds of 30 to 40 miles per hour while also offering much greater reliability, comfort and safety than horse-drawn carriages. This made it feasible to travel hundreds of miles in a single day, and to rapidly move freight and mail from one place to another.[13]

Today we think of railroads as continuous long-distance lines of steel rails over which powerful locomotives drag enormous loads of freight and passenger cars at moderate to high speeds. But the basic idea underlying them dates back many centuries, at least to ancient Greece where tracks were cut into limestone so that wheeled vehicles could reliably be hauled from one side of the Corinth peninsula to the other. By the fourteenth century, this basic idea had been revived in central Europe as a way of transporting coal and minerals up out of the depths of mines. The floors of mine tunnels were rarely smooth, and there was no effective way to pave them. Instead, carefully spaced wooden *rails* were laid down, and wheeled tubs or carts were then pushed or pulled along them bearing heavy loads of ore. By the seventeenth century, this idea had been extended to simple above ground *wagon-ways*. These featured wooden rails over which horse-drawn wagons could transport coal from mines to nearby canals where it would be loaded into barges. By the 1760s, plates of cast iron were being laid over the wooden rails to provide greater durability, and by the early 1800s solid cast iron rails started to be used on some wagon-ways.[14]

As Boulton and Watt successfully marketed their improved reciprocating steam engines in the 1770s—followed in the 1780s by models capable of producing rotary motion—a number of inventors began to wonder whether these new power plants might be combined with existing wagon-ways to create steam-powered land transportation. A number of attempts were made, although none proved to be successful primarily because of the great weight of the Watt engine. Both for technical and business reasons, Boulton and Watt favored designs that made their steam engines large and heavy. In particular, they didn't believe that the boilers then available were strong enough to resist the force of steam under anything much greater than atmospheric pressure, and for this reason their power plants featured heavy oversized pistons and cylinders able to accommodate large amounts of low-pressure steam. On ships, this great weight was still workable. But for land vehicles, such massive engines were a tremendous barrier, since they had to be powerful enough to move not only their own weight, but that of any carriages, freight and passengers.

Watt's concern about iron boilers exploding under the force of high-pressure steam was not unfounded. Iron makers in the eighteenth century had difficulty controlling the quality of the iron plates they produced, and on a number of occasions exploding boilers proved themselves capable of killing people and causing great property damage. By the final decades of the eighteenth century, however, as rapid industrialization created ever-increasing demand for iron products, new methods were found to make tougher and more consistent wrought-iron plates, the best of which were strong enough to reliably contain high-pressure steam. In fact, one person who was intensely interested in utilizing high pressure, or *strong* steam was Watt's own collaborator William Murdoch (or Murdock; 1754–1839).[15]

Starting as a young employee of the firm, Murdoch had developed some of the company's most notable inventions including the sun and planetary gear transmission. He was a brilliant innovator and inventor in numerous fields, and he later became well known for pioneering the use of illuminating gas starting in the 1790s. In the 1780s, however, while still working for Boulton and Watt, Murdoch designed and built a working model of a road wagon, or steam car, with an engine employing steam at above atmospheric pressure, a design that achieved a significantly higher ratio of power-to-weight than typical steam engines. On several occasions, Murdoch successfully demonstrated it indoors—it was fueled by alcohol—and he tried hard to get Boulton and Watt interested in developing both the high-pressure engine concept and the road wagon. The two older men, however, though extremely talented engineers and manufacturers, were also relatively conservative businessmen who were highly protective of their market dominance and not always open to innovation. As a result, they refused to use the resources of the firm to develop Murdoch's new technology, and in fact seem to have done everything possible to discourage Murdoch from pursuing the idea. Secretly, however, Watt added language to a steam engine patent application that he filed to cover its use in road wagons.[16]

Boulton and Watt missed the opportunity to become pioneers in high-pressure steam engines, although their firm eventually went on to build them. Yet Murdoch's innovation still ended up playing a role in the development of this next major advance in artificial power. In the mid-1790s, a young Cornish mechanic and inventor named Richard Trevithick (1771–1833) is believed to have witnessed a demonstration of Murdoch's model steam carriage, and late in that decade, the two also seem to have been neighbors for a time. Trevithick had grown up near the mines in Cornwall—his father had been a mine operator, and his father-in-law had started a successful iron foundry which specialized in making the steam engines intended to operate mine pumps. Around 1797—in the same period when he and Murdoch were neighbors—Trevithick began modifying a steam engine that was installed at the mine where he worked. His overall goal was to create a much lighter machine working at several times atmospheric pressure which could produce more power in spite of having a smaller piston and cylinder. First, he developed a double-acting cylinder design that allowed the steam to move the piston both up and back in the cylinder. Next he fitted the cylinder with improved steam valves. Then, he decided to do away with Watt's condensing chamber by releasing the spent steam directly into the atmosphere after transferring some of its heat to the cold water being let in to the boiler. This might sound wasteful until one remembers that the change also eliminated the considerable weight of both the condensing chamber and the large reservoir of cold water needed to cool it. In the years that followed, Trevithick further improved his design by replacing the heavy masonry firebox with a much lighter iron fire chamber located inside the shell of the boiler.[17]

These innovations, taken together, resulted in the first really workable steam engine running at higher than atmospheric pressure, and as predicted it produced far more power for its weight than did a typical Boulton and Watt engine. It was not really light in comparison to say a small, modern automobile engine, but it weighed so much less than a conventional steam engine that Trevithick soon resolved to pursue Murdoch's idea of using it to power a road wagon. Three years later, on December 27, 1801, his first steam powered land vehicle, the first *horseless carriage*, ran successfully under its own power, although it soon suffered structural damage due to the roughness of the local roads. Its usefulness at the time was severely limited by the lack of smooth, paved road surfaces, but it demonstrated to Trevithick the potential of high-pressure steam, and two years later, he built a second steam-powered carriage which he ran successfully over an eight-mile route in London.[18] (Murdoch had designed and built only a working scale model before being dissuaded from pursuing it further by Watt and Boulton, and the steam military wagon built by Nicolas-Joseph Cugnot in 1769 was not a practical device.)

Richard Trevithick had many ups and downs in his career, and in retrospect it seems that he was just a few years too early for the development of really workable steam railroads. High-pressure steam, though, did not fall by the wayside, and in 1812 Matthew Murray designed and built two high-pressure cog-driven steam colliery locomotives which successfully hauled coal over a route that came to be known as the Middleton Steam Railway. Not long after, several similar locomotives were placed in service in other coal yards.[19]

The next person to dramatically advance railroad technology was the British civil and mechanical engineer George Stephenson. Around 1813 he saw one of Matthew Murray's locomotives working in a colliery near Newcastle, and resolved to construct his own version. The next year, he built the *Blucher* to pull coal trains in a colliery in nearby Killingworth, and then—joined by his son Robert— continued to work on locomotive design over the next fifteen years. During this period, the pair continued to improve the design of high-pressure engines to yield more and more power. Around the same time, iron manufacturers figured out how to *hot roll* railroad rails in order to produce metal with the strength and flexibility of wrought iron. This allowed the replacement of the earlier brittle cast iron rails. As a result, Stephenson was able to build large, powerful locomotives that could dispense with the cog wheels on Murray's colliery locomotives and rely instead on their own weight to provide adequate traction even when running slightly uphill.[20]

By 1825, Stephenson's company had started building locomotives for a new industrial railroad, the 25-mile-long Stockton and Darlington line, which was intended to connect a number of different coal mines with the port of Stockton in northern England. The line opened on September 27, 1825, and as part of the festivities, 600 passengers were slowly hauled a distance of 12 miles, offering a convincing demonstration of the potential of passenger railways. Four years later, Stephenson went on to design another new locomotive named *Rocket* for a planned railway linking the port city of Liverpool and the growing manufacturing center of Manchester. He then entered it into a competition known as the Rainhill Trials, and *Rocket* was the only locomotive among the five that competed that actually managed to complete the full trial—at an average speed of 12 miles per hour. Stephenson received the prize of five hundred pounds and even more importantly won the contract to supply engines for the new Liverpool and Manchester railway. The line opened the following year, and Stephenson's name soon became synonymous with the development of railway transportation not only in Britain but in other countries as well.

Railways, Coal and Carbon Dioxide Pollution

At the opening of the Liverpool and Manchester Railway in 1830, Britain had less than 100 miles of steam railway. Just fifteen years later, however, railway mileage had grown more than twenty-fold to over 2,200 miles. Seven years after that, around 1852, it had increased to roughly 7,000 miles—fully a seventy-fold increase in twenty-two years. Such rapid expansion was revolutionary, and yet in the next forty-eight years to 1900, British railway mileage more than tripled again to roughly 22,000 miles.

In 1838, there were 5.5 million passenger trips on Britain's young rail network. Just seven years later, in 1845, thirty million trips occurred annually, and by 1855, more than 110 million passenger trips were recorded around the country. Britain's railroad system was the first to rapidly expand, but a similar pattern was soon seen in other industrializing countries. In Germany, railroad mileage went from zero in 1830 to 35,000 miles (60,000 km) in 1912; in the United States, it went from 40 miles of track in 1830 to 163,000 miles (262,000 km) just sixty years later.[21] Wherever tracks were laid and regular train service established, steam-powered railways transformed overland transportation and revolutionized society. Goods, people, mail and news could now move safely and inexpensively over long distances while the traditional isolation of inland places became significantly reduced wherever new rail lines were built.

The expansion of railroads and gradual replacement of sailing vessels by steamships created a demand for large amounts of fuel, and this, in turn, led to tremendous increases in the mining and burning of coal. This can most easily be tracked in the statistics available for Great Britain.[22] In 1700, shortly before Newcomen's first pumping engine was built, it is estimated that total British coal production was roughly three million tons and then took seventy-five years to double to six million tons by 1775. (In the 1700s, coal was mostly used for space heating, iron-smelting, salt-making, brewing, brick-making and pumping water out of mines.) Forty years later, in 1815, as Boulton–Watt engines were becoming more common in factories across the country, coal use had risen nearly three-fold again to 16 million tons.

The 1830s and 1840s saw the first great flush of expansion in Britain's railways, and by the 1850s, the output of British coal is estimated to have amounted to fully 72 million tonnes (figures from 1850 on are given in metric tonnes, a unit equal to 1,000 kilograms or 1.102 tons). Thereafter, with industrialization and railway expansion in full swing, it took only twenty-five years for production to once again double to 140 million tonnes by the 1870s. And thirty-five years later, in the first decade of the twentieth century, the United Kingdom contained more than 3,200 deep mines producing coal, and the total coal being dug each year reached roughly 260 million tonnes. (By that time, electrification was also underway, at least in cities, and some coal had begun to be used in power plants to generate electricity.) In roughly two centuries, British coal production had increased more than 85-fold.

The burning of coal produces a wide range of waste products (see Chapter 7), but one of the most important from a long-term environmental point of view is carbon dioxide (CO_2). Much of the mass of coal is carbon, the ancient remains of the carbon dioxide that was captured by plants using photosynthesis millions of years ago. But the amount of carbon in coal from different sources varies considerably, from less than 70 percent in the lowest grades of bituminous coal, to over 90 percent in anthracite coal. When a carbon atom is burned, it combines with two atoms of oxygen, each of which weighs more than the carbon atom itself. For this reason, burning just one and a quarter pounds of coal containing one pound of carbon will end up releasing 3.667 pounds of carbon dioxide into the atmosphere.[23]

Around 1905, when Britain annually produced 260 million tonnes of coal, its burning generated more than 625 million tonnes of carbon dioxide (this assumes that the coal burned was on average 80 percent carbon). Over the course of the nineteenth century, other countries had also expanded their coal production, in some cases even faster than Great Britain. America's production in 1905, for example, was one and a half times as great as Britain's, and this was at a time when Britain was producing roughly a quarter of the world's coal.[24] This meant that worldwide, more than two and a half billion tonnes of carbon dioxide was being released into the atmosphere from coal burning each year. This upward trend continued throughout the twentieth century, and by 2010, world production of coal had reached more than *seven billion tonnes*, at least *seven times* the amount that had been produced a century earlier. By that point, global atmospheric releases of CO_2 from coal burning had reached the staggering figure of nearly 18 billion tonnes, or about fifty-five percent of the total CO_2 being emitted. (In 2010, total CO_2 releases reached an estimated 33 billion tonnes *from all sources*.)[25]

Electricity

Early Discoveries

People's earliest experiences with electricity are lost in pre-history, although as far back as 2,500 years ago, a Greek named Thales of Miletos wrote down some early observations. Thales rubbed chunks of amber (fossilized conifer resin) with pieces of cloth or animal fur and found that for a short while the rubbed amber was able to attract other lightweight objects. What he had found was *static electricity*, a build-up of electrical charge on the surface of an object. (The English word electricity, which first began to be used in the 1600s, comes from a Latin term which means *from amber*.) Thales was unable to explain what he observed, and so static electricity remained just a curiosity. Two millennia later, as the European scientific revolution began to gather momentum, static electricity once again became a topic of interest, and the next significant advance occurred in the mid-1740s when German and Dutch researchers independently developed the *Leyden Jar*, a device for storing an electrostatic charge. They published their design and the device was soon adopted by experimenters. Among them was the American, Benjamin Franklin, whose subsequent experiments with electricity led to the development of the first practical *lightning rod*, although neither Franklin nor anyone else in his era was getting close to understanding the underlying nature of electricity itself.[26]

Advances in Theory and Practical Technology

Around 1820, a series of discoveries finally began to lay the groundwork for transforming electricity from a laboratory curiosity with very few practical applications into an essential component of technology. In April 1819, Hans Christian Ørsted (1777–1851), a professor of physics at the University of Copenhagen, was setting up some apparatus for a class lecture when he observed that a flow of electricity in a wire caused a nearby compass needle to deflect. He also noticed that when he turned the power off, the needle quickly swung back to its original position. Then he tried re-connecting the wire so that the current flowed in the opposite direction. When he did this, the compass needle once again deflected, but in the *opposite direction*. From this, Ørsted concluded that there was a relationship between magnetism and electricity. Soon, the problem was taken up by Andre-Marie Ampere (1775–1836), a French physicist, who showed that when the current in two electrified wires ran in opposite directions, the wires bent away, but when the current ran in the same direction, the two bent towards each other.[27]

Ampere duly published his results, and together with Ørsted's they stimulated a brilliant series of experiments by a young and largely self-educated English physicist and chemist named Michael Faraday (1791–1867). In 1821 Faraday carried out two different critical demonstrations on electricity's relationship to magnetism. In one of these, he ran a current through a fixed wire around which a magnet was free to turn. When he turned on the current, the magnet began to spin; but when he turned off the current, it stopped. This showed that the flow of electricity was generating a magnetic field that could interact with the field produced by the permanent magnet. Another of his experiments involved a fixed magnet around which a wire was free to rotate. When the wire was electrified it began to rotate around the magnet, once again showing that the current was producing a magnetic field that could interact with the magnet's own permanent field.[28]

These experiments led Faraday to support Ørsted's conclusion that electricity and magnetism were closely related phenomena. (Note how dramatically the pace of scientific advance was accelerating at this time. Important experiments, publications, and further discoveries were made by three different researchers in three different countries in only three years, even before telegraphs, telephones and railroads were available to speed up the spread of information.) Two years later, another English scientist named William Sturgeon invented the first *electromagnet* by wrapping un-insulated wire around an iron bar that had been varnished to insulate it electrically. When the current was turned on, the device became so powerfully magnetized that it was able to lift many times its own weight.

After that, experimentation on electricity and magnetism quickened even further. In 1827, a German physicist, Georg Simon Ohm, made a careful study of the flow of electricity through wires of different lengths and thicknesses, establishing two important principles: thicker wires were able to conduct more electricity with less losses than thinner wires; and as wires became longer and longer, they lost more and more of the electricity flowing through them. (These discoveries eventually became part of *Ohm's Law* of electrical resistance.) In 1831 Faraday performed an experiment in which a single iron ring was wrapped over part of its surface with one insulated conducting wire, while a second insulated wire was wrapped around a different part of the same ring. When he ran current through the first wire, he discovered that electricity also began to flow through the second wire, even though the two were not physically connected. To explain this, he reasoned that the flow of electricity in the first wire had generated a strong magnetic field that was able to pass right through the other wire's insulation and *induce* a flow of electricity in it. This was a startling discovery that confirmed his model of electromagnetism as involving invisible lines of force that spread out in all directions from magnets and electrified wires. Not long after, it came to form the basis for electrical *transformers*, key devices that were able to change the electrical voltage (or pressure) of a flow of current.

Another of Faraday's critical discoveries was that moving a wire or other conductor inside a magnetic field caused a current to flow in it. To demonstrate this, in 1831 he devised an insulated copper wheel whose edge spun inside the arms of a large horseshoe magnet. As the wheel turned, an electric current developed in it which could then be passed out through a wire to power devices normally run by chemical batteries. This device was the first general purpose *electrical generator,* and his research also suggested how the generator's opposite, the *electric motor,* could be constructed. Later that same year, the American physicist Joseph Henry did construct the first electric motor, although although some believe that others did it before him. Faraday eventually summarized the key principles he had discovered about electromagnetism in a formulation we know today as *Faraday's law of induction.* In everyday language, it basically says that changing magnetic fields are able to generate electric currents.[29]

Faraday and other physicists continued to experiment with various aspects of electricity and magnetism over the next several decades. In the mid-1830s, for example, Joseph Henry went on to invent the electric relay, a kind of switch that could be activated by a weak current over a relatively long wire to operate a strong electromagnet at a distance. By the 1840s, this device, combined with more reliable electricity from improved batteries, had led to the creation of successful electric telegraphs, the world's first modern communications technology. Telegraph lines soon spread rapidly throughout the industrialized world allowing messages to reach their destinations thousands of times faster than in the past. By 1861, a line had been built clear across North America from coast to coast, making possible delivery in minutes of transcontinental messages that earlier would have taken weeks to arrive. By 1867, an even longer line was completed right across Eurasia, allowing telegrams to fly back and forth between locations as far apart as England and India.[30]

The middle decades of the nineteenth century marked the beginnings of modern electrical technology, its growth spurred by the increasing number of university-based research physicists, many of whom experimented extensively with various aspects of electricity, magnetism and light. (In the 1840s, for example, Faraday had shown that a magnetic field could change the polarization of light, indicating that light and magnetism were also related phenomena.) By the end of the 1850s, university scholars and non-academic inventors had amassed a large collection of experimental results and working principles related to electricity laying the groundwork for a deeper, scientific understanding of the phenomenon. In 1861 and 1862 this led to one of the most important breakthroughs in nineteenth-century physics, when the Scottish scientist James Clerk Maxwell synthesized a large mass of observations and measurements into a single set of rigorous mathematical rules governing electricity, magnetism and related phenomena.[31]

Maxwell was one of the nineteenth century's leading theoretical physicists as well as being a highly gifted mathematician. His papers systematized electromagnetism and laid out the precise mathematical relationships governing electrical, magnetic and optical phenomena. He confirmed the conjectures by Faraday and others that electricity, magnetism and light were all closely related phenomena, grouping them together as different types of *electromagnetic radiation*. Maxwell's unification of these seemingly disparate kinds of energy was also a key step in the overall development of modern science, and helped to lay the foundation not only for Albert Einstein's work on relativity, but also for much of modern physics.[32] In addition, Maxwell's work provided a solid, quantitative basis for the newly emerging field of electrical engineering, and this, in turn, allowed the safe and reliable design of an endless array of electrical (and later electronic) devices beginning with the commercial electrical generator, the telephone and the local electric power grid.

Electrification

As we saw in Chapter 4, by the early 1800s many cities and towns in northwestern Europe had constructed networks of pipes to supply coal gas for illumination, marking a dramatic improvement over the candles, torches and oil lamps that had earlier been relied on for artificial light. Gaslight, however, still left a great deal to be desired. Its brightness was limited; lamps needed maintenance; and there was an ever-present danger of fire, explosion or asphyxiation.[33] Early nineteenth-century experiments like those of Humphry Davy with electric arcs had shown that electricity could be used for lighting, and later researchers also found that electricity could be made to produce light inside an evacuated glass tube.

By the 1840s, as batteries improved and the number of scientists and inventors continued to increase, serious efforts got underway to produce light using electricity. The two greatest challenges were finding materials for the electrified filaments that could stand up to the intense heat produced, and developing manufacturing techniques that could cheaply and reliably produce and seal the high vacuums needed for the bulbs. There were many experimenters and many small advances, but perhaps the leading figure in the light bulb's early development was the British physicist Joseph Swan. He started working on it around 1850, and by 1860 had produced a demonstration model. At that point, though, he turned to other work because at that time neither adequate supplies of electricity, nor equipment to reliably create high vacuum inside bulbs was available. Fifteen years later, improvement in vacuum pumps and the adaptation of reciprocating steam engines to power generation led him to resume his work, and by 1878 he had succeeded in demonstrating a workable light bulb for which he received a British patent in 1880.[34]

On the other side of the Atlantic, Swan's work was being closely followed by the pioneering American electrical inventor and businessman Thomas Edison. In 1878 he began to work intensively on further improving Swan's bulb, although according to some sources, the bulb that Edison eventually developed was largely a copy of the Englishman's design. In any case, Edison succeeded in obtaining a US patent for his device.[35] As he worked on light bulbs, Edison was also getting ready to market entire electrical systems to communities, including generators powered by reciprocating steam engines, distribution lines, light fixtures, switches, bulbs, etc. In 1880, in fact, Edison's company patented a complete electrical distribution system, and in 1882 he installed two of these local systems, one in New York and one in London. Five years later, 121 Edison power stations had been built in the United States. Edison's system utilized direct current (DC) electricity, the same type produced by batteries. It was well suited to light bulbs and the electric motors of the day, but also had important limitations. At moderate voltage, DC current could not be sent more than a mile without rapidly being lost from the wires, and this meant that Edison's generators were restricted to locations in central business districts where customers were very close by.[36]

Soon, however, a rival system utilizing alternating current (AC) had been designed by Nikola Tesla, a Serbian engineer who had briefly worked for Edison but had never been able to convince him of AC's advantages. After leaving Edison's company, Tesla teamed up with George Westinghouse, the inventor of the railway air brake, to develop and market AC electrical systems. Edison's patents had given him a virtual monopoly on DC systems, and for several decades his company fought a bitter campaign against the implementation of AC systems. (It also created an advertising campaign in the 1880s claiming that Edison, not Joseph Swan, was the inventor of the first modern incandescent light.) The advantages of AC were so great, however, especially the ease with which its current could be stepped up to high voltage, sent long distances through relatively thin wires, and then stepped back down to safe working voltages, that in the twentieth century it was almost universally adopted.[37]

A Breakthrough in Power Generation

After Maxwell's work, waves of inventors labored to produce useful new electrical devices, while chemists and metallurgists created many new industrial materials that could be used in electrical components (see Chapter 4). In the mid-nineteenth century, however, steam engines still remained the only practical artificial power source apart from those limited locales where hydroelectric power was available. The alternative—dependence on cumbersome, low-powered chemical batteries—offered very limited potential to expand the uses of electricity, and as time went on, it became clear that there was a great need to more efficiently harness the power of steam to generate electricity. Faraday had shown that electricity could be generated using rotating devices, and ever since the Boulton–Watt era at the turn of the nineteenth century, reciprocating steam engines had been able to produce rotary motion, most notably in railway locomotives and in engines driving line shafts in factories. But despite Edison and Swan's exciting, pioneering electrical installations which used reciprocating steam engines to run electrical generators (dynamos), power generation by steam engines met with only limited success, primarily because of steam engines' low operating speed and limited efficiency.

Over the course of the nineteenth century, the efficiency of piston steam engines had very slowly increased as improved boiler and condenser designs allowed steam to be generated and handled at ever higher pressures and temperatures. But even after these advances, they were still only able to transform about 15 percent of the total energy released in their boilers into useful work—85 percent of the power in the coal or wood had to be thrown away as unrecoverable waste heat.[38] They also operated at relatively low speed due to the massive nature of their large pistons and cylinders while efficient electrical generators needed to have their shafts turned steadily at speeds of 1,000 revolutions per minute (rpm) or more. In addition, piston steam engines contained many moving parts which were subject to wear and break down.

As early as the 1840s, the search for greater efficiency had led steam engineers to experiment with rotary steam engine designs—steam turbines—although for a long time, no one had been able to overcome the difficulties involved in designing, manufacturing and operating such a device. Designs based on low-pressure steam simply didn't generate enough power, and their relatively low operating temperatures dramatically limited their potential overall efficiency. High-pressure steam, on the other hand, was harder to generate and handle, and it hit the turbine with tremendous force creating serious vibration and sometimes even doing physical damage to the turbine's blades. For these reasons, efficient, workable steam turbines remained beyond the reach of engineers throughout most of the nineteenth century.

In the 1880s, however, this problem was taken up by a young Anglo-Irish engineer named Charles Algernon Parsons. Parsons, who like Maxwell, was extremely gifted in mathematics, was the son of a famous British astronomer, and also had a hands-on bent towards practical engineering. In addition, he was familiar with the advanced water wheels with curved blades known as water turbines that had marked a significant improvement over flat blade water wheels (see Chapter 8). Parsons reasoned that since steam under pressure behaved somewhat like pressurized water, a steam turbine might be designed as a series of modified water turbines. Determined to use super-hot, high-pressure steam in his designs in order to make his engines powerful enough, Parsons decided that he could reduce physical damage and reach higher efficiencies if his turbines extracted energy from the steam in a series of small steps, rather than in a single ferocious blast. To achieve this, he created a multi-rotor or compound design in which the first part of the turbine had small blades that extracted only a fraction of the steam's energy, before passing it on to the next section that had slightly larger blades. The second set extracted another fraction, before passing it on to the third set of slightly larger blades, and so on. This process continued through several different sections, each with larger and larger blades designed to extract similar amounts of energy from the steam as it progressively fell in pressure and temperature. A great deal of mathematical analysis was involved in developing the exact design. Parsons had to carefully balance the resistance of the rotor's blades in a given turbine section against the energy contained in a given quantity of steam at a particular temperature and pressure. But he was up to the task.[39]

Besides the enormous challenges of turbine design, Parsons had other problems to overcome, particularly the large in-line (axial) and centrifugal forces generated by rapidly spinning turbines. The axial thrusts tended to wear out the shaft bearings, while the centrifugal forces could make turbines vibrate dangerously. Normal engineering practice of the day would have called for massive anchoring bases to avoid destructive movement. Parsons, however, dealt with the in-line forces in a more creative and ingenious way by dividing the turbine into two sections and bringing the steam in through the middle so that the two halves generated axial thrusts in opposite directions, canceling one another out. His solution to the centrifugal force problem grew out of the basic multi-step turbine design, which allowed both the central shaft and the steam itself to move more slowly than would have been possible in a single-stage power plant. Having solved these daunting engineering problems, Parson's turbine paved the way for a revolutionary increase in electricity production. His very first working turbine, built in 1884, was immediately connected to a small electric dynamo that produced 7,500 watts. But his firm and others worked hard over succeeding decades to improve efficiency and power, particularly through improvements to the design of the turbine blades and the steam nozzles. By 1899, a turbine generator capable of generating fully 1,000,000 watts had been placed into service.[40]

Turbine steam generators soon became the standard for electric-power generation around the world, a role they continue to play up to the present day. By the year 2000, steam-driven turbines were producing more than 80 percent of the world's electricity, with the steam to run them being produced through the burning of coal, oil, gas or biomass, or by the heat of nuclear fission. To achieve higher and higher efficiencies, fairly complex designs and very high temperatures and pressures were being used. Some current turbines are designed to run on steam heated to more than 1,100°F (610°C) at pressures as high as 3,700 pounds per square inch—260 times atmospheric pressure.[41] They also develop a high level of vacuum in their condensers, and this helps to rapidly cool the spent steam that is turning back into water, increasing the overall efficiency of the turbine's cycle.

Modern steam turbines re-cycle their steam both within the turbine itself and within the overall installation. One reason for this is that they have to start with expensive, ultra-pure water in order to avoid the build-up of mineral deposits inside the turbine which might otherwise damage the blades or partially block the passageways. They also divert small amounts of their steam at different points in the overall cycle to heat inlet water and other parts of the machinery that must be kept hot. All of this goes on hidden inside the tens of thousands of power plants large and small that have been built around the world.[42] Most people are completely unaware of the details of this central technology of modern civilization. Yet almost everyone alive depends upon it, since the power it produces runs our homes, our offices, our factories, our hospitals, our communications networks, our militaries and everything else powered by grid electricity. Once steam-driven units had been developed for electrical generation, the basic technology of turbines able to run at very high temperatures also proved to have other important uses. By the 1930s, they had been adapted to run directly on the hot gases produced by burning fossil fuels rather than steam, and this opened up two new important applications: direct gas-fired turbines for generating electricity and turbo-jet aircraft engines.

Light bulbs were only the first of hundreds of familiar devices and appliances powered by electricity that came to be used in homes, offices, factories, motor vehicles, ships, planes and everywhere else: fans, heaters, refrigerators, radios, televisions, microwave ovens, computers, medical equipment, and on and on. Some, like the electric bread toaster, may not have represented a really significant change in the way that people lived. But others like the washing machine and vacuum cleaner dramatically reduced the heavy burdens of domestic work, while electrically powered telephones, radios, televisions and computers transformed social and political life by ushering in an era of rapid business, personal and mass communication. Grid electricity also spurred enormous innovations in electric motors and their controls creating a gradual revolution in industry as belt-driven tools powered off central shafts run by steam engines were little by little replaced with individual machines powered by their own electric motors. Even public transportation became electrified, with streetcars and urban train systems rapidly giving up steam engines in favor of cleaner and more reliable electric power. Electric lighting also transformed workplaces, homes, offices and schools, as adequate levels of illumination could now be provided regardless of the season, the weather, or the cycles of day and night.

Little by little, over the course of the twentieth century, centrally produced electricity became available to the majority of people in high and middle-income countries, and also to some in poorer nations. As electrification spread, demand for power grew continuously and with it the need to utilize larger and larger amounts of fuel to power the steam turbines and run the generators (or to dam more rivers to produce hydroelectricity). In addition, large investments were needed in local and regional electrical grids in order to distribute the power. Creating them involved production and installation of vast networks of electrical cables joining large numbers of transformer stations and widely dispersed power plants. This also meant that large amounts of copper had to be mined and refined for all the cables and wires needed, processes which, in turn, required large amounts of energy and also produced significant pollution. In the early twentieth century, new synthetic chemicals also came to be used in some components of electrical grids, especially in transformers and capacitors. Often, these were insulated with PCBs (poly-chlorinated biphenyls), substances that later turned out to be highly persistent environmental pollutants (see Chapter 4).

Until recent decades, regular statistics on electric power production and fossil fuel consumption were not generally kept on a national or global level. For Great Britain, however, reliable information describing the mix of fuels used to generate electricity is available reaching back to about 1920. (To allow different fuels to be compared, all statistics are quoted in units equal to *one million tonnes of oil equivalent*.) In 1920, total British fuel use in power plants equaled 3.68 units, more than 99 percent of which was coal. By 1930, it had risen more than 60 percent to 5.78 units. By 1940 it was 10.72 units, and by 1950 it had nearly doubled again to 19.06 units. Throughout the rest of the twentieth century and into the early twenty-first, the energy Britain used to generate electricity climbed steadily, reaching a peak of 87.06 million tonnes (equivalent) in 2005 before declining a bit as programs of conservation, efficiency improvements and renewable sources began to have an effect. In 2005, the fuel consumed to generate electricity was twenty-three times greater than it had been in 1920. From the 1960s on, British nuclear power stations made an increasing contribution to power generation displacing 23 million tonnes of fossil fuels at their peak output in 1998, after which nuclear's share slowly dropped. The total amount of coal burned increased steadily until the 1980s, peaking at 51.58 million tonnes in 1987 before significantly declining as large supplies of natural gas became available. (In 2012, however, coal's share of the fuel mixed increased again due to economic forces.) In 2012, 40 percent of British greenhouse gas emissions came from the production of electricity. Since total UK CO_2 output that year was around 480 million tonnes, power production alone accounted for over 190 million tonnes of carbon dioxide.[43]

Power from Petroleum

The early days of modern petroleum technology are clouded by a variety of conflicting claims about who was the first to achieve different milestones. What is known, however, is that by the late 1840s and 1850s recognizably modern oil wells had begun to be drilled in Asia, Europe and North America, and that the most important commercial product derived from them was kerosene, which could be used as lamp oil. By the 1860s, steam engines were being used to power the drilling operations. At the time, many other crude oil components enjoyed much more limited market demand. A great deal of the lighter fractions and fuel gases, which today are so highly valued for gasoline and cooking, were worth almost nothing back then because there were no workable engines that could utilize the liquids, and no infrastructure existed for transporting or compressing the gas.

Petroleum, as we saw in Chapter 4, is an extremely complex mixture that must undergo extensive chemical separation, reformation and purification before highly usable products are produced. A few of its most familiar components are light gases like methane and propane; light liquids like hexane and octane (used in gasoline); medium-weight liquids containing hydrocarbons like nonane (9 carbons), decane, undecane, dodecane, etc., up to hexadecane (16 carbons) that are useful for kerosene, diesel and heating oil; heavier liquids commonly burned in ships and sold as residual oil; waxes like paraffin; and very heavy tars and asphalts. Other materials like benzene and sulfur are also reasonably common in petroleum, although nowadays refineries try to remove them. If not removed, benzene increases the risks of cancer from exposure to the unburned fuel; sulfur, if burned along with the hydrocarbons, creates irritating air pollution and acid rain. Throughout the age of the automobile, refiners have tended to place particular emphasis on maximizing the output of motor fuels from each barrel of crude oil.

Internal Combustion

Steam engines were *external combustion* devices—the fires that ran them were in their boilers not in their power cylinders. For useful work to be accomplished, energy had to be transferred to their pistons in the form of energetic steam. Even before Thomas Newcomen's early pumping engine, the obvious limitations of external combustion had led some researchers to suggest harnessing power from something resembling controlled cannon fire. In 1678, for example, the Dutch scientist Christiaan Huygens had designed a device run by gun powder in which an explosion in a modified cannon would be used to raise a heavy piston. As the piston fell back down due to gravity, its force could be used to lift heavy objects or to pump water. It is believed that he actually made and tested a prototype, but it is doubtful that it would have proved practical at the time given the limitations of manufacturing and machining technologies.[44] And once Thomas Newcomen had introduced his pumping engine in the second decade of the eighteenth century, experimenters tended to focus on steam rather than on explosives as a new source of power.

Significant progress on *internal combustion* did not really begin until the middle of the nineteenth century when some inventors began using flammable gases to fuel their experimental engines, although the un-pressurized gases they tried didn't produce much power when burned. Refined petroleum products, in contrast, once they became available, offered a much more promising source of power. Gasoline, for example, contained enough energy to provide considerable power, yet as a liquid it was still easy to pump into a combustion chamber and to mix with air. It also burned with little or no ash and produced combustion wastes that could readily be vented to the outside.

Serious inventions based on internal combustion began around 1860 when the Belgian Jean Etienne Lenoir demonstrated a workable engine fueled by coal gas, and employed it to power a primitive *horseless carriage*. Most of the units that Lenoir's company actually delivered, however, were stationary engines used to power water pumps or machines such as printing presses. Also in the 1860s, the German Nikolaus Otto began developing internal combustion engines of several different designs intended to run on volatile, light density hydrocarbons like gasoline. By 1872, two other talented German engineers, Gottlieb Daimler and Wilhelm Maybach, had begun to collaborate with Otto, and in 1876 Otto developed his famous four-stroke gasoline engine which is still the most commonly used power plant in automobiles around the world. In the late 1870s and early 1880s, yet another German engineer named Karl Benz independently developed both two-stroke and four-stroke engines (using Otto's power cycle), and by 1885 Benz had started to manufacture and sell early automobiles.[45]

Making working horseless carriages, however, required much more than simply an engine and a wagon. As Benz's company worked to develop practical vehicles, it also had to develop or improve a whole series of important automobile components which today are taken for granted: a mechanism for generating high voltage pulses of electricity to create the ignition sparks; durable electrodes (spark plugs) to ignite the fuel; a special rechargeable battery for starting the engine; an improved carburetor for mixing fuel and air; a gear-box, gear shift and clutch to allow the engine to always remain within a reasonable range of speeds; and a water-filled radiator to dissipate excess engine heat.[46] Even then, practical automobiles would not have been possible without other important nineteenth-century inventions including pneumatic tires and the vulcanized rubber needed to make them.

Compression-Ignition Engines

During the same decades when many breakthroughs were being made in engineering automobile components, other European engineers were working hard on internal combustion engines that could burn heavier oils instead of light, gasoline-like mixtures. Quite a few individuals made contributions, but the two most notable ones were those of the British inventor Herbert Akroyd Stuart and the Franco-German engineer Rudolf Diesel (1858–1913).

Rudolf Diesel, the inventor of the engine that bears his name, was born to German parents in Paris in 1858; in 1870 he was sent to Germany to complete his education, and there he did well in school and soon began to study engineering. One of his teachers in Munich was Carl von Linde who pioneered gas separation and refrigeration technologies (the two are related), and then went on to establish a company that even today is a leading supplier of industrial gases. After completing his first engineering degree, Diesel started working for Linde who was then setting up and running a large refrigeration plant in Paris. Diesel also studied the theoretical analysis of *heat engines* written in the 1820s by the French military engineer and physicist Sadi Carnot. Carnot had shown that engine efficiency was related to the temperature difference between the expanding gases (steam, burning gasoline, etc.) which pushed the piston and the much cooler air outside the device. As the temperature of combustion increased, so did the potential efficiency of the engine. Diesel also understood from his work on the separation and compression of gases that air itself, if adequately compressed, could get hot enough to cause ignition. With these principles in mind, he turned his attention to designing an engine that would be able to ignite its fuel without electric sparks.[47]

Diesel was aware that the gasoline-burning Otto cycle engine also compressed its fuel–air mixture to a fraction of its starting volume, improving the mixing of the air molecules with the fuel molecules, and bringing them physically closer together. This also raised the mixture's temperature and improved the engine's thermal efficiency. In engines burning gasoline, however, there was a serious limit to how much the fuel–air mixture could be squeezed. If compressed too much, gasoline would get so hot that it would start to burn before the piston had come to the top of the cylinder or the spark plug had fired. In modern American usage this is called *pre-ignition*, and it sets a practical limit of about 10:1 on how much the fuel–air mixture in a gasoline engine could be shrunk in volume. If squeezed more than that, the spontaneous combustion that resulted would seriously disrupt the engine's timing and power production. Diesel knew that the ignition temperature for heavier oils was much higher than for gasoline, and he reasoned that they could undergo greater compression without undesirable pre-ignition.[48]

Taking all these considerations into account, Diesel set out to design an engine whose compression ratio and cycle starting temperature would be higher than those of gasoline engines, typically as high as 15:1 or even 20:1, with the compressed air reaching 550°C or 1,022°F. At such temperatures, the air would spontaneously ignite any fuel that it came into contact with, making it possible to do away with the complexities of the gasoline engine's electrical ignition system. In his design, the fuel could be introduced into the combustion chamber just at the instant when the piston was squeezing the air to its smallest volume and highest temperature. And there would no longer be a need for expensive, hard-to-refine gasoline: his device could run on a relatively heavy oil that could be distilled directly from petroleum. To contain such high compression, the construction of the engines had to be somewhat sturdier and more massive than gasoline ones, but they could also be somewhat simpler, even though the traditional carburetor did have to be replaced with a pressurized fuel pump. Because the highly compressed air got very hot, and because diesel fuel was heavier than gasoline and so generated larger amounts of burning gas, compression-ignition engines also turned out to be notably more energy efficient than those which burned gasoline.[49]

The generation to which Diesel belonged marked a turning point not just in the groundbreaking inventions they managed to develop, but also in the way that leading technologists came to be trained. Up to that point, most nineteenth-century engine developers had been trial and error engineers, albeit ones with some ability to quantitatively analyze their designs and prototypes. Diesel, in contrast, was a scientifically trained engineer, in full command of both the relevant physical theories and the mathematical techniques necessary to develop his designs. In 1893—the same year that he was granted a patent for his compression-ignition system—he also published an important book entitled *Theory and Construction of a Rational Heat-engine to Replace the Steam Engine and Combustion Engines Known Today.* He seems to have understood exactly what he was doing, and rather than progressing mostly by tinkering, as even James Watt had done, his design was developed primarily through theory. His first working model was produced in 1897, and almost immediately he began licensing his design to manufacturers. By 1903, engines similar to Diesel's had begun to be installed in ships; by the 1920s they were powering trucks; by 1930 farm tractors; and by 1933 passenger automobiles. Despite being relatively noisy and producing significant pollution, they proved to be reliable, efficient and relatively easy to maintain. In the twenty-first century, significantly re-engineered compression-ignition engines now offer relatively clean-burning, highly fuel-efficient power plants that are being used in passenger automobiles and trucks.[50]

The Rise of Automobiles

The last four decades of the nineteenth century saw extraordinary progress in power technologies including internal combustion engines. Yet even after all the breakthroughs of Lenoir, Otto, Benz and Diesel, most cars were still expensive, high-tech curiosities that were not practical replacements for railroads or old-fashioned horses. Most petroleum was still being rather crudely refined into kerosene for lamps, and significant quantities of gasoline were not yet widely available. Unlike coal-fired steam railroads that had offered reliable high speed transportation for decades, most people saw automobiles as experimental technologies, or toys for the rich.

By the early years of the twentieth century, however, all this quickly began to change. After 1900 car production rapidly began to increase as standardization of parts and more rapid means of assembly replaced the *one-at-a-time* hand-crafted approach to making automobiles. This pattern can most clearly be seen in statistics for the United States, which had the most dispersed rural population of any fully industrialized country, and hence represented the greatest potential market for automobiles. In 1903 only 13,000 cars were produced in America, but just four years later almost three times as many were made—36,000. Four years after that, in 1911, US production had again more than quadrupled to 167,000. By 1915, largely due to the onset of mass production at Ford, it quadrupled again to more than 800,000 cars a year. In 1916, US production passed one million vehicles, and by 1919 it stood at roughly 1.4 million. In 1923, Ford alone produced nearly two million vehicles, and total US production exceeded 3.1 million.[51] In only two decades, from 1903 to 1923, the number of American automobiles *produced annually* had increased 238-fold!

The economic boom years of the 1920s were followed by the Depression era 1930s and the war years of the 1940s, when US production of civilian automobiles dropped significantly (although production of military motor vehicles soared dramatically).[52] By 1950, however, civilian production had once again taken a sharp turn upwards, with fully eight million autos being produced in America. In 1975, nearly 9 million autos were manufactured, and in 2000, the peak year so far of US production, more than 12 and a half million cars and light trucks were made. By the end of the twentieth century, however, the United States was producing only a small fraction of the world's autos, and in 2000, other nations produced more than 45 million vehicles, raising global production to 58 million. By 2012—largely as a result of dramatically increased Chinese output—world production shot up again to 84 million per year.[53] Most cars, of course, last for years, and by 2010 it was estimated that there were more than one billion (1,000,000,000) motor vehicles in use around the world.[54]

What does it take to run one billion autos? Putting aside the energy and materials needed to make them, the main thing they consume, once built, is fuel, primarily gasoline and diesel, which are both made from petroleum. As motor vehicles use soared during the twentieth century, so did the extraction, refining and burning of petroleum, although some petroleum products were also used to run ships, generate electricity, heat homes and make new materials like plastics. The United States was an early leader in oil extraction and refining as well as in automobile production; its pattern of oil use during the twentieth century illustrates the challenges that power technologies have created for environmental stability.

In 1920, when annual US auto production was already about two million vehicles, American crude oil production reached one million barrels (42 million gallons) per day for the first time. By 1950, when four times as many autos were being produced each year, oil extraction was five times as great (210 million gallons daily). Twenty years later, in 1970, even though automobile production had not significantly increased over 1950, accelerated oil exploration and pumping had brought US production to a peak of nearly ten million barrels a day.[55] From 1970 on, depletion of readily recoverable reserves caused US domestic oil production to fall, even as US car production continued to rise. (In the twenty-first century oil production began to rise again due to new petroleum extraction techniques.) US petroleum consumption, however, continued to increase as more and more oil came to be imported from other countries, rising from 714 million gallons a day in 1980 to roughly 880 million gallons a day in 2005. In that year, domestic production was only 218 million gallons a day which meant that roughly 660 million gallons a day had to be imported in order to meet the vast US domestic demand.[56]

By the end of the twentieth century, the US had also been joined by numerous other countries in its consumption of large quantities of petroleum products. By 2012, worldwide oil consumption reached a staggering 3.8 billion gallons daily (roughly 90 million barrels).[57] In that year, the total of all the oil burned around the world is estimated to have produced nearly 11 *billion* tonnes of carbon dioxide, or slightly more than one third of the 30 billion tonnes of CO_2 that entered the atmosphere that year. (Burning coal and peat release even more CO_2 than oil, while combustion of natural gas produces somewhat less; intentional burning of forests also contributes an important fraction.)

Other Environmental Effects of Automobiles

In the nineteenth century, before the development of automobiles, most land transportation involved either recently built railways or traditional travel by foot, horse or wagon. Wagons were generally made of wood, a largely renewable resource, and draft animals caused little or no pollution. Trains, in contrast, were noisy and often spewed heavy smoke and soot into the air from their coal boilers. In a few areas, the building of railway lines also led to extensive forest clearance in order to supply railroad ties. Overall, however, railways did not cause truly massive environmental damage.

Automobiles had a very different environmental footprint, partly because cars and trucks spread out over the landscape in a way that railroads never did. For this reason, they caused some environmental damage just about everywhere that people went, even along some otherwise uninhabited road corridors that were built to connect far-flung settlements. Cars required quantities of steel and other metals to build and wire them, and plastics for interior and exterior trim. Like trains they also caused significant air pollution and noise as they burned their petroleum fuels. They also came to have a significant content of hazardous materials as various inventions were added to make them faster, more reliable and more capable. Hydraulic brakes, for example, needed special fluids to transfer force from the pedal to the brake pads, and after trying many different materials, those that worked best at reasonable cost turned out to be ones that contained toxic substances. Later, when automatic transmissions were introduced, their internal fluids added an additional poisonous component. These fluids posed no hazard to driver or passenger while they were safely contained within a vehicle, but during servicing, or at the end of a car's life, leakages into the environment sometimes occurred. Stopping a car or truck also involved generating a lot of friction in a controlled way, and this was the job of the brake shoes or pads and the drums or disks that they pushed against. The drums were steel, but the pads were designed not to slip even when intense friction heated them up, and so for decades they were made partly from heat-resistant, but toxic, asbestos. The pads and shoes were designed to gradually wear away, revealing fresh friction surfaces, and as they did so, some of the toxic asbestos was slowly released into the environment. Brake repair personnel were also sometimes exposed to high levels of asbestos while doing their work.[58]

In addition to cars needing brakes and automatic transmissions, their engines also needed a way to get rid of waste heat. Generally this was accomplished by circulating a liquid coolant through the engine block and then out to a radiator. It was possible to use straight water as coolant, but in freezing weather, ice could ruin an engine, and there were also limitations to water's thermal performance. Early automobiles usually mixed toxic wood alcohol with water, but the former had a tendency to boil away. Finally, in the middle of the twentieth century, somewhat toxic ethylene glycol mixed with water came to be the standard coolant in cars, even though its use is slowly being replaced by non-toxic propylene glycol.[59]

These were not the only harmful materials that came to be commonly used as car components. Batteries contained large amounts of lead, and in the early decades of the twentieth century soluble lead compounds also came to be added to gasoline in order to smooth combustion (see Chapter 4). These two sources created significant lead pollution which impacted some children's neurological development and also caused other environmental problems. Over time, engine oil becomes dirty and loses some of its lubricating ability so that it has to be replaced. For decades, many car owners just poured used oil onto the ground or down a sewer where it contaminated ground or surface water. By the second half of the twentieth century, scientists had also discovered that such used motor oil contained combustion products that could contribute to the development of cancer.

After early experiments in the 1930s, auto manufacturers began to offer air conditioning units on new cars, and by the 1980s, the majority of new cars sold in the US included them.[60] Most used chlorofluorocarbon compounds (CFCs) as refrigerants, compounds like di-chlorodifluoromethane, which had one carbon atom bonded to two chlorine and two fluorine atoms. Auto air conditioners had to operate in high temperature environments inside engine compartments where they were also subject to continual mechanical shocks. As a result, they were much more likely to leak refrigerant than were stationary cooling units. Up until the 1990s, these widely used compounds were also commonly released into the atmosphere whenever auto air conditioners and refrigeration and freezing units were serviced.

In 1985, however, scientific research published in the journal *Nature* showed that the chlorine from escaped refrigerants was causing a significant thinning of earth's protective high-altitude ozone layer, leading to increased danger of skin cancer and other biological damage, including possible harm to plankton populations in the sea. In response, a movement got underway to phase out chlorofluorocarbons and replace them with something less ozone-depleting, the most readily available substitutes being compounds in which a chlorine atom was replaced by a hydrogen atom. After significant negotiations, a highly successful international treaty known as the Montreal Protocol agreed to a gradual phase-out of the most chlorine-rich refrigerant molecules, and a gradual phase-in of the less heavily chlorinated refrigerants known as hydrochlorofluorocarbons (HCFCs). By the second decade of the twenty-first century, this had led to a decrease in the area of ozone holes of roughly 25 percent over its historical maximum extent, although as of 2012 it still covered an area equal in size to all of North America plus Mexico.[61]

HCFCs, however, still contained some chlorine, and their continuing use was not likely to produce a final solution to ozone depletion, so later revisions to the Montreal Protocol called for their replacement by compounds known as hydrofluorocarbons (HFCs) containing only hydrogen, carbon and fluorine. Most new refrigeration units now use them, although recent research has shown that all three families of refrigerant chemicals, including the planned HFCs, are also extremely powerful greenhouse gases that contribute to global warming. Of the three, the HFCs, which are least damaging to the ozone layer, are also among the most potent warming gases known. Their molecules have the potential to trap thousands of times as much heat as molecules of carbon dioxide and they remain present in the atmosphere for decades or even centuries after release. In the fall of 2016, the Montreal Protocol was up-dated at a meeting in Kigali, Rwanda and a new target was set to start phasing out refrigerants which either cause global warming or harm the ozone layer by 2028. In the twenty-first century, the overall use of refrigerants is increasing dramatically as more and more people around the world become able to afford the purchase of refrigerators and air conditioners (stationary and automobile). Unfortunately, the search for an efficient, environmentally sound, cost-effective refrigerant is still on-going, and in the meantime both the protective ozone layer and the global heat balance remain somewhat at risk.[62]

So far we have described some of the direct environmental impacts of automobiles.[63] But once large numbers of fast-moving cars had replaced horses and horse-drawn vehicles, un-paved roads also increasingly became a problem. Dirt roads almost always eroded to the point of being very rough to drive on, and in heavy rains they tended to wash out. In dry weather, rapidly moving cars and trucks also threw up dangerous amounts of dust. The solution, beginning in the nineteenth century, was to construct smooth, waterproof road surfaces, initially by coating them with oil or paving them with some kind of concrete. Plain oil, however, had to repeatedly be renewed and could be slippery in wet weather, while typical concrete made with Portland cement was strong, but also expensive, hard to smooth, and liable to cracking. As different paving materials were tried, it was found that a type of concrete bound together with asphalt rather than cement had the best combination of properties for road use—smoothness, durability and cost.

As we saw in Chapter 4, asphalt (or bitumen) is a major component of the residues remaining after crude oil has been fractionally distilled. It is heavy, sticky, resistant to bacterial attack, and very waterproof, and so is an excellent material for binding road paving together. Chemically speaking, it is also rich in a class of compounds known as polycyclic aromatic hydrocarbons, some of which, in isolation, are toxic or cancer-causing. But because they are also very insoluble in water and mostly are encapsulated in the pavement itself, they seem to pose relatively little danger to people or to the environment unless they are burned or used in water pipes. Sometimes, shredded up old tires (crumb rubber) are also added to asphalt pavement mixes, and when this is done, there is some potential for water pollution from leaching of hazardous compounds present in the rubber. This practice does however provide an environmentally sound way of re-cycling some of the millions of tires that are discarded each year. Also on the plus side, after decades of treating old asphalt paving as trash or fill, by the end of the twentieth century, it had come to be one of the most completely recycled materials in a number of industrial countries. Crushed, and then mixed in with new asphalt paving materials at ratios up to 1:1, it has performed well in many places. As a result, in some places up to 99 percent of old ground pavement is recycled.[64]

By the mid-twentieth century, asphalt pavement had been widely adopted and had come to dominate road paving around the world. In the United States by 2011 there were more than *two and a half million miles* of paved roads, the vast majority of which were surfaced with asphalt, not counting large additional areas of private driveways and parking lots.[65]

To date, asphalt paving has not been found to produce significant water or air pollution, and the fact that it can repeatedly be recycled offers a distinct advantage compared to many other construction materials. On the other hand, it does require some continual supply of virgin material from petroleum, the production of which does have environmental impacts. Its generally dark color can also make heavily paved urban areas hotter than they normally would be, and its waterproof surface, like other paving and roofing materials, tends to increase rapid storm water runoff and flooding risk. (Calcium hydroxide can be pressed into the upper surface of newly laid asphalt to change its color to light gray which reduces its heat absorption from the sun.[66])

In the last century or so, ever-improving cars and trucks and more and more paved highways have led to vastly increased automobile travel at higher and higher speeds. In the U.S. by the early 2010s, people traveled more than three *trillion* (3,000,000,000,000) miles every year in cars and trucks (254 million registered vehicles in the US, each driven an average of more than 13,000 miles per year).[67] Roads were built where it was convenient for people. Often they ended up divided natural habitats or interfering with wildlife travel corridors with the result that many millions of mammals, birds and reptiles are killed every year by autos, with some human deaths also occurring when large mammals like deer, moose or elk hit vehicles as they attempt to cross roads.[68] In some locations such as Banff National Park in Canada and the Florida Panther National Wildlife Refuge in the United States, extensive runs of fencing, combined with wildlife tunnels, bridges and one-way gates have been installed to reduce automobile–wildlife collisions, although such measures are expensive and require continual maintenance.

Around the world, people have come to take petroleum-burning motor vehicles for granted. Yet they are a very mixed technology from an environmental point of view. In addition to noise pollution, they cause damage and environmental disruption at every phase of their life cycle, from the extraction of metal ores and petroleum required to make them, to the air pollution and hazardous material leaks associated with running them, the toxic substances utilized in some of their components and the difficulties in disassembling and recycling them at the end of their useful life.[69] Electric cars, once they become widely accepted, will dramatically reduce the air pollution from automobile operation. But most of the other environmental impacts associated with cars and the roads they run on will still remain.

Aviation

Along with the railway and the automobile, the nineteenth century also saw renewed interest in another innovative means of transportation—air travel. An early pioneer was the British engineer George Cayley, who laid down a theoretical framework for flying machines around 1800 and then went on to build a number of successful piloted gliders over roughly five decades.[70] Towards the end of the nineteenth century, the German Otto Lilienthal became the leading experimenter with piloted gliders; he made over 2,000 flights in the 1890s before perishing in a crash. His well-publicized work was a great inspiration to later aviation pioneers.[71]

Powered flight had to wait for the development of small internal-combustion gasoline engines in the last decades of the nineteenth century. By 1903, after developing some of the earliest tools of modern aeronautical engineering including a wind tunnel to test their designs, America's Wright brothers were able to repeatedly demonstrate controlled, powered flight. These early machines, however, were fragile and in need of constant maintenance, and so airplanes remained mostly experimental for the rest of that decade. After 1910, however, as compact gasoline engines became more and more powerful, attempts were made to use airplanes to deliver the mail and to transport passengers. In the 1920s, following numerous technical advances and the much publicized exploits of military pilots in World War I, many more airplanes started to be built, and both long-distance postal services—*airmail*—and regular passenger flights began to be offered. Airplane designers and manufacturers were also busy, and that decade also saw the introduction of the first passenger planes capable of carrying more than a handful of people. Among these were France's Farman Goliath, a three engine model built by Fokker and Ford, and a range of *flying boats* designed to take off and land on water. By the 1930s, recognizably modern (i.e., markedly streamlined, metal-skinned and with retractable landing gear) passenger planes had begun to appear including Boeing's ten-passenger 247 (1933), and the Douglas DC-3 (1936), which was able to carry up to 32 passengers. The 1930s, of course, also brought the Great Depression which reduced levels of economic activity and growth in air travel, and from the mid-1930s through the mid-1940s, many countries were also involved in preparations for, or the actual battles of World War II. Many advances in aviation continued to be made in this period, although there was a heavy emphasis on military applications.[72]

Throughout aviation's first three decades, airplanes and their engines had grown steadily larger and more powerful. By the 1920s, however, engineers had come to realize that there were limitations to the amount of power that propellers could provide, and so they began to look for other possible ways of propelling aircraft. They were aware of rockets, but because they had to carry both the weight of their fuel and of an oxidizing agent, they were not highly efficient. Nor were rockets well-suited to hours of continuous cruising, or to operations like landing.

One person who thought long and hard about this problem was a young English aviation engineer named Frank Whittle, who in the late 1920s was slowly progressing through a series of Royal Air Force (RAF) training courses. Two ideas that appealed to him were to have a traditional piston propeller engine include an air compressor for better combustion and also to have the exhaust gases directed out behind to provide more thrust. Others had earlier speculated about such an engine, although it still didn't deal with the limitations inherent in propellers. Finally, Whittle had the radical idea of doing away with the propeller altogether so that the propulsion would come entirely from the blast of hot exhaust gases. Just as Charles Parson had radically re-designed the steam engine, Whittle now called for an aircraft engine based on a liquid-fueled turbine that compressed its own air and directed a massive backward-blasting, un-muffled flame through an accelerating nozzle to push the plane forward. Even at high altitudes in thin air, he argued, such a compressor could provide enough oxygen to fully burn all the fuel, maximizing the engine's power and efficiency.[73]

Whittle had outlined the design for the world's first *turbojet*, and after laying out its main features, he tried hard in 1929 to get his own RAF to develop it. The RAF was an aeronautical technology leader in those early days and had an active program to create advanced aircraft. But the *highly qualified* expert to whom they sent Whittle's design for review did not believe that it would work and rejected it. Whittle was discouraged, but he decided to go ahead and patent the new engine design anyway, something he would have been barred from doing if the RAF had decided to adopt it for development. By 1935, he had joined with others to set up a small private company to continue working on it, although limited funding and on-going lack of support from his own government made the engine's development painfully slow. Finally, in late 1940, his first test engine was run successfully, and by May 15, 1941, the first full-scale test flight was successfully completed.[74]

While Whittle was working on his pioneering jet engine in Britain in the mid-1930s, another talented aeronautical engineer named Hans von Ohain was independently developing a similar design in Germany. The Nazi Government, which was rapidly developing its armed forces, quickly took up the concept, assigning von Ohain to work with a group of experienced aircraft engineers to develop a jet engine for use in military fighter aircraft. Provided with ample government resources, they made rapid progress and were able to stage the world's first successful jet-plane test flight in August 1939. By 1941, the Germans had succeeded in producing the world's first operational jet fighter, the Messerschmitt Me-262. (Britain had jet fighters in advanced testing that year, and the United States by 1942.) Despite these engineering successes, however, it ended up taking several additional years for these countries to produce significant quantities of the new planes, and their eventual introduction into combat in 1944 ended up having little impact on the outcome of the war.[75]

Following World War II, jet engines came increasingly to be used for military aircraft and soon evolved into several different distinct types. Most notably, straight turbojets came to be replaced in the 1960s with turbofan engines which soon became the standard for both military and civilian planes. Turbojets derived all of their thrust from their exhaust; turbofan engines, in contrast, used their turbines to power both an air compressor and a huge fan that forcefully blasted air backwards adding extra thrust to the exhaust gases. Turbofans were also quieter and more efficient.[76]

The first commercial passenger jet, the de Havilland Comet, was introduced in 1952. It was followed by the Boeing 707 in 1957, and the Douglas DC-8 in 1959. Airlines soon had their choice of reliable, comfortable aircraft, and in the 1960s and 1970s, with generally good economic times around the world, commercial jet travel exploded. In the United States, for example, in 1960 there were roughly 60,000,000 individual one-way plane trips and the total mileage all passenger traveled equaled 1,180,000,000 miles (one billion one hundred and eighty million). Most of those flights were still being made in propeller planes powered by piston or turboprop engines. Twenty-five years later, in 1985, people took more than six times as many plane trips—380,000,000—and traveled more than twelve times as far—over 14,000,000,000 (fourteen billion) miles. And twenty-five years after that, in 2010, there were 786,000,000 trips averaging 1,353 miles apiece and covering the astonishing total distance of over one trillion passenger miles (1,063,458,000,000).[77] Over those same decades, advances in engine and airplane design did manage to achieve meaningful reductions in the fuel needed to move one passenger one mile, but even these very important efficiency gains were vastly overwhelmed by the roughly 900-fold increase in total passenger miles flown.[78]

By 1985, world consumption of jet fuel—a blended, mixed-weight, flammable oil similar to kerosene—had risen to 1,902,000 barrels per day (79,884,000 gallons or 302,360,000 liters).[79] When burned, it produced 764,496,000 kilograms (1,685,700,000 pounds or 764,496 metric tonnes) of carbon dioxide. By 2008, this had increased to 5,269,000 barrels (221,298,000 gallons) a day. Jet airliners had become more fuel efficient per passenger mile, but the constant blast of planet-warming carbon dioxide pollution had nonetheless more than doubled (roughly 9.57 kilos of CO_2 are produced per gallon of jet fuel burned[80]).

In coming decades, air travel is expected to increase at a rapid rate, with some experts predicting that it will double again over the next twenty years. At the same time, there is little doubt that jet airliners will continue to become somewhat more efficient, especially as metal plane bodies start to be replaced with lighter composite materials like carbon fiber reinforced polymer. Efficiency gains, on average, are *declining*, however, and it seems highly unlikely that technological advances will permit significant increases—let alone a doubling—of air travel with no additional production of waste carbon dioxide. If, of course, it turned out that increased passenger air mileage led to comparable reductions in travel by automobile or other greenhouse gas producing activities, more and longer plane trips would not be as great a problem. But that has not been the pattern over recent decades, and instead, the number of automobiles in use has continued to climb. In the future, it seems likely that people will end up traveling more total miles in both cars and planes. And if this happens, greatly increased carbon dioxide pollution leading to accelerated global warming will be nearly impossible to avoid.

Notes

1 WaterHistory.org, Water wheels, www.waterhistory.org/histories/waterwheels; N. Rodgers, 2008, *Roman Empire*, Metro Books; Wikipedia, Water wheel, https://en.wikipedia.org/wiki/Water_wheel.

2 NNDB, Denis Papin, www.nndb.com/people/558/000096270.

3 J. S. Allen, 2004, Thomas Newcomen, *Oxford Dictionary of National Biography*, www.oxforddnb.com; Wikipedia, Thomas Newcomen, https://en.wikipedia.org/wiki/Thomas_Newcomen.

4 C. Lira, 2013, Brief history of the steam engine, www.egr.msu.edu/~lira/supp/steam; Wikipedia, Newcomen steam engine, https://en.wikipedia.org/wiki/Newcomen_atmospheric_engine.

5 Wikipedia, History of the steam engine, https://en.wikipedia.org/wiki/History_of_the_steam_engine.

6 J. Tann, 2004, James Watt, *Oxford Dictionary of National Biography*, www.oxforddnb.com.

7 C. Lira, 2013, Brief history of the steam engine, www.egr.msu.edu/~lira/supp/steam.

8 J. Tann, 2004, James Watt, *Oxford Dictionary of National Biography*, www.oxforddnb.com.

9 Industrialrevolution.sea.ca, The Industrial Revolution, http://industrialrevolution.sea.ca/innovations.html; D. Hunt, 2004, Edmund Cartwright, *Oxford Dictionary of National Biography*, www.oxforddnb.com.

10 Wikipedia, Steamboat, https://en.wikipedia.org/wiki/Steamboat.

11 US Army Corps of Engineers, A history of steamboats, www.sam.usace.army.mil/Portals/46/docs/recreation/OP-CO/montgomery/pdfs/10thand11th/ahistoryofsteamboats.pdf.

12 Wikipedia, Steamboat, https://en.wikipedia.org/wiki/Steamboat.

13 Jane Austen's World, The difficulties of travel and transportation in early 19th C. Britain, https://janeaustensworld.wordpress.com/2012/04/25/the-difficulties-of-travel-and-transportation-in-early-19th-c-britain.

14 Wikipedia, History of rail transport, https://en.wikipedia.org/wiki/History_of_rail_transport.

15 C. Lira, 2013, Brief history of the steam engine, www.egr.msu.edu/~lira/supp/steam.

16 BBC, William Murdock's steam locomotive, www.bbc.co.uk/ahistoryoftheworld/objects/1h1Eko8qR_WF9MhWcySpuA.

17 P. Payton, 2004, Richard Trevithick, *Oxford Dictionary of National Biography*, www.oxforddnb.com.

18 T. Ricci, 2012, Richard Trevithick, www.asme.org/engineering-topics/articles/transportation/richard-trevithick.

19 Grace's Guide to British Industrial History, Middleton Colliery Railway, www.gracesguide.co.uk/Middleton_Colliery_Railway; G. Cookson, 2004, Matthew Murray, *Oxford Dictionary of National Biography*, www.oxforddnb.com.

20 M. W. Kirby, 2004, George Stephenson, *Oxford Dictionary of National Biography*, www.oxforddnb.com; Wikipedia, George Stephenson, https://en.wikipedia.org/wiki/George_Stephenson.

21 A Web of English History, The growth and impact of railways, www.historyhome.co.uk/peel/railways/railways.htm; Wikipedia, History of rail transport in Germany, https://en.wikipedia.org/wiki/History_of_rail_transport_in_Germany; Wikipedia, History of rail transport, https://en.wikipedia.org/wiki/History_of_rail_transport.

22 UK Department of Energy and Climate Change, Historical coal data: coal production, availability and consumption 1853 to 2012, www.gov.uk/government/statistical-datasets/historical-coal-data-coal-production-availability-and-consumption-1853-to-2011; Wikipedia, History of coal mining, https://en.wikipedia.org/wiki/History_of_coal_mining.

23 Taftan Data, Combustion of coal, www.taftan.com/xl/combus1.htm.

24 Wikipedia, History of coal mining, https://en.wikipedia.org/wiki/History_of_coal_mining.

25 Netherlands Environmental Assessment Agency, 2011, Long-term trend in global CO_2 emissions, report) www.pbl.nl/en/publications/2011/long-term-trend-in-global-co2-emissions-2011-report.

26 I. Asimov, 1972, *Asimov's Guide to Science*, Basic Books.

27 R. D. A. Martins, Resistance to the discovery of electromagnetism: Orsted and the symmetry of the magnetic field, http://ppp.unipv.it/Collana/Pages/Libri/Saggi/Volta%20and%20the%20History%20of%20Electricity/V&H%20Sect3/V&H%20245-265.pdf.

28 F. A. J. L. James, 2011, Michael Faraday, *Oxford Dictionary of National Biography*, www.oxforddnb.com; Wikipedia, Michael Faraday, https://en.wikipedia.org/wiki/Michael_Faraday.

29 electrical4u.com, Faraday Law of Electromagnetic Induction, www.electrical4u.com/faraday-law-of-electromagnetic-induction; M. Doppelbauer, Karlsruhe Institute of Technology, The invention of the electric motor 1800–1854, www.eti.kit.edu/english/1376.php.

30 Wikipedia, Telegraphy, https://en.wikipedia.org/wiki/Telegraphy.

31 Engineering and Technology History Wiki, James Clerk Maxwell, http://ethw.org/James_Clerk_Maxwell; Wikipedia, James Clerk Maxwell, https://en.wikipedia.org/wiki/James_Clerk_Maxwell.

32 P. M. Harman, 2009, James Clerk Maxwell, *Oxford Dictionary of National Biography*, www.oxforddnb.com.

33 Scientific American, 1869, Illuminating gas—what it is, and how it is made, *Scientific American* (March 20), www.scientificamerican.com/article/illuminating-gas-what-it-is-and-how.

34 F. Andrews, A short history of electric light, www.debook.com/Bulbs/LB01swan.htm; C. N. Brown, 2011, Sir Joseph Wilson Swan, *Oxford Dictionary of National Biography*, www.oxforddnb.com.

35 The Historical Archive, The history of electricity—a timeline, www.thehistoricalarchive.com happenings/57/the-history-of-electricity-a-timeline.

36 Wikipedia, Thomas Edison, https://en.wikipedia.org/wiki/Thomas_Edison; V. Smil, 2005, *Creating the 20th Century*, MIT Press.

37 I. Asimov, 1972, *Asimov's Guide to Science*, Basic Books.

38 H. W. Dickinson, 1963, *A Short History of the Steam Engine*, Frank Cass.

39 C. A. Parsons, 1911, The steam turbine, https://en.wikisource.org/wiki/The_Steam_Turbine.

40 C. Gibb and A. McConnell, 2008, Sir Charles Algernon Parsons, *Oxford Dictionary of National Biography*, www.oxforddnb.com.

41 GE Power Generation, Combined cycle steam turbines fact sheet, https://powergen.gepower.com/content/dam/gepower-pgdp/global/en_US/documents/product/steam%20turbines/Fact%20Sheet/steam-turbines-fact-sheet-dec-2015.pdf.

42 Wikipedia, List of power stations, https://en.wikipedia.org/wiki/List_of_power_stations.

43 UK Department of Energy and Climate Change, 2013, UK greenhouse gas emissions, www.gov.uk/government/uploads/system/uploads/attachment_data/file/193414/280313_ghg_national_statistics_release_2012_provisional.pdf; UK Department of Energy and Climate Change, Historical electricity data: 1920–2014, www.gov.uk/government/statistical-data-sets/historical-electricity-data-1920-to-2011.

44 R. L. Galloway, 1881, *The Steam Engine and its Inventors: A Historical Sketch*, Macmillan; Wikipedia, Gunpowder engine, https://en.wikipedia.org/wiki/Gunpowder_engine.

45 New World Encyclopedia, Internal combustion engine, www.newworldencyclopedia.org/entry/Internal_combustion_engine.

46 Wikipedia, Karl Benz, https://en.wikipedia.org/wiki/Karl_Benz.

47 Wikipedia, Rudolph Diesel, https://en.wikipedia.org/wiki/Rudolf_Diesel.

48 Wikipedia, Diesel engine, https://en.wikipedia.org/wiki/Diesel_engine.

49 Busch-Sulzer Bros, 1913, *The Diesel Engine*, http://books.google.com/books?id=FEV-AAAAIAAJ&printsec=frontcover&dq=diesel+engine&hl=en#v=onepage&q&f=false.

50 Diesel Technology Forum, What is clean diesel?, www.dieselforum.org/about-clean-diesel/what-is-clean-diesel.

51 Wikipedia, US automobile production figures, https://en.wikipedia.org/wiki/U.S._Automobile_Production_Figures.

52 The National World War II Museum New Orleans, By the numbers: wartime production, www.nationalww2museum.org/learn/education/for-students/ww2-history/ww2-by-the-numbers/wartime-production.html.

53 Wikipedia, Automotive industry, https://en.wikipedia.org/wiki/Automotive_industry.

54 Wards Auto, 2011, World vehicle population tops 1 billion units, http://wardsauto.com/ar/world_vehicle_population_110815.

55 US Energy Information Administration, US field production of crude oil, www.eia.gov/dnav/pet/hist/LeafHandler.ashx?n=PET&s=MCRFPUS1&f=A.

56 Index Mundi, United States crude oil consumption by year, www.indexmundi.com/energy.aspx?country=us&product=oil&graph=consumption.

57 US Energy Information Administration, International energy statistics, www.eia.gov/cfapps/ipdbproject/iedindex3.cfm?tid=5&pid=alltypes&aid=2&cid=ww,&syid=2006&eyid=2010&unit=TBPD.

58 Ontario Ministry of Labor, Alert: asbestos hazard in vehicle brake repair, www.labour.gov.on.ca/english/hs/pubs/alerts/i34.php.

59 Wikipedia, Ethylene glycol poisoning, https://en.wikipedia.org/wiki/Ethylene_glycol_poisoning.

60 M. S. Bhatti, 1999, Riding in comfort II: evolution of automotive air conditioning, *ASHRAE Journal* (September), www.ashrae.org/File%20Library/docLib/Public/2003627102420_326.pdf.

61 US National Oceanographic and Atmospheric Administration, 2012, Antarctic ozone hole second smallest in twenty years, www.noaanews.noaa.gov/stories2012/20121024_antarcticozonehole.html.

62 USEPA, Ozone layer protection, www.epa.gov/ozone-layer-protection/international-actions-montreal-protocol-substances-deplete-ozone-layer.

63 USEPA, Region 10, Environmental impacts from automobiles, http://yosemite.epa.gov/r10/owcm.nsf/Product+Stewardship/autos-impacts.

64 Clemson University, Department of Civil Engineering, Benefits of bubberized asphalt, www.clemson.edu/ces/arts/benefitsofRA.html.

65 US Federal Highway Administration, Public road length, www.fhwa.dot.gov/policyinformation/statistics/2011/hm12.cfm; National Asphalt Pavement Association, History of asphalt, www.asphaltpavement.org/index.php?option=com_content&view=article&id=21&Itemid=41.

66 J. J. Emery et al., 2014, Light-coloured grey asphalt pavements: from theory to practice, *International Journal of Pavement Engineering* 15(1): 23–35.

67 US Federal Highway Administration, Average annual miles per driver by age group, www.fhwa.dot.gov/ohim/onh00/bar8.htm.

68 J. Davenport and J. L. Davenport (eds.), 2006, *The Ecology of Transportation: Managing Mobility for the Environment*, Springer.

69 US EPA Region 10, Environmental impacts from automobiles, http://yosemite.epa.gov/r10/owcm.nsf/Product+Stewardship/autos-impacts.

70 J. A. Bagley, 2006, Sir George Cayley, *Oxford Dictionary of National Biography*, www.oxforddnb.com.

71 Flying Machines, Otto Lilienthal, http://flyingmachines.org/lilthl.html.

72 Avjobs, History of aviation—first flights, www.avjobs.com/history.

73 G. B. R. Feilden, 2008, Sir Frank Whittle, *Oxford Dictionary of National Biography*, www.oxforddnb.com.

74 Ibid.
75 Century of Flight, Hans von Ohain, www.century-of-flight.net/Aviation%20history/jet%20age/Hans%20von%20Ohain.htm.
76 National Aeronautics and Space Administration, Turbofan engine, www.grc.nasa.gov/WWW/k-12/airplane/aturbf.html.
77 US Bureau of Transportation Statistics, Airline passengers, flights, freight and other air traffic data, www.rita.dot.gov/bts/data_and_statistics/by_mode/airline_and_airports/airline_passengers.html; US Bureau of Transportation Statistics, Historical air traffic statistics, annual 1954–1980, www.rita.dot.gov/bts/sites/rita.dot.gov.bts/files/subject_areas/airline_information/air_carrier_traffic_statistics/airtraffic/annual/1954_1980.html.
78 P. M. Peeters et al., 2005, Fuel efficiency of commercial aircraft, www.transportenvironment.org/sites/te/files/media/2005-12_nlr_aviation_fuel_efficiency.pdf.
79 Index Mundi, World jet fuel consumption by year, www.indexmundi.com/energy.aspx?product=jet-fuel&graph=consumption.
80 US Energy Information Agency, Fuel emissions coefficients, www.eia.gov/oiaf/1605/coefficients.html.

7

ENVIRONMENTAL IMPACTS OF FOSSIL FUELS

Coal

In recent years, it has become possible to work out the various kinds of damage that different technologies impose on the living world. But it is worth keeping in mind that a comprehensive scientific understanding of the natural environment only emerged in the second half of the twentieth century, and that back in the eighteenth and nineteenth centuries there was relatively little insight into the complex and often irreversible destruction that human activities could cause. It is also important to remember that over the thousands of years of agricultural civilization preceding industrialization, chronic overpopulation leading to frequent episodes of devastating famine and disease had produced a psychology of grim desperation in our ancestors who often had to live in a permanent state of insecurity about whether they would be struck down by famine or disease. Over time, this state of mind became a deep-seated cultural value that made it seem almost natural to overlook impacts such as the disappearance of native plant and animal populations, the wastage of topsoil due to over-cropping and over-grazing, or the massive pollution of water, air and soil.

As we saw in Chapter 4, coal, oil and gas are all produced over millions of years through gigantic geological *composting* processes, as enormous layers of organic matter buried under sediments away from earth's atmosphere are slowly transformed, first by microorganisms, and then by heat, pressure and chemical change. Of the three fuels, coal is the most abundant, and also the easiest to find and use, and it was the first to be widely exploited. Unlike some mined substances, however, coal is not a very uniform material; rather it consists of a large family of organic, sedimentary rocks whose composition in any given bed varies according to the dead organic starting materials, the conditions of its burial, and the length of time it has been entombed. The earliest stage of coal formation is peat which usually consists of at least two thirds recognizable dead plant material.[1] In northwestern Europe, for example, bog peat has been harvested for fuel for thousands of years. It is smoky when burned, but still has been widely used for space heating and cooking, and in the twentieth century, a few peat-burning electric power plants were also built in places like Ireland where peat was abundant.

After peat, the next stage of coal formation is brown coal or lignite. It is considered the poorest fuel grade of coal because it contains more water and mineral matter and so releases less heat when burned. It also produces significant air pollution and large amounts of ash. Coal that has been further transformed, especially by pressure, takes the form of bituminous coal. Though it burns hotter than brown coal and hence makes a better fuel with less total ash, many of the organic compounds it originally contained have been transformed into bitumen, a tarry substance similar to the asphalt that is derived from petroleum. Asphalt, as we have seen, does have commercial applications in road pavement and roof shingles, but it is not usually economical to extract it from coal. For this reason, bituminous coal is usually burned as is—bitumen and all—leading to the production of considerable smoke and soot.[2]

Coal that has been further transformed beneath the earth's surface takes the form of anthracite, a dark, shiny, rock-like material, consisting almost entirely of carbon. Anthracite burns cleanly with a high heat and produces very little ash. It is the scarcest but most sought-after type of coal. Around the world, the limited supplies of it are used for power generation, metallurgy, and space heating. Lastly, if coal is buried long enough and deep enough, every chemical element in it save one will separate out. When this happens, the material that remains is graphite which is pure carbon. Graphite occurs in small but economically significant deposits, and is most familiar to us as the black substance used in pencil lead.

By the eighteenth century, people had already been using coal for thousands of years for heating, cooking and refining metals. In regions like Europe, more and more had come to be dug, transported and burned as growing populations and increasing deforestation had produced ever greater shortages of firewood. Over the course of the eighteenth, nineteenth and twentieth centuries, its utilization increased even faster, producing both great economic benefit and considerable hard to the environment. One way to understand its overall environmental impact is to trace its path from the mines where it is dug to the final wastes that remain after it is burned.

Coal Mining

Here and there on the earth's surface, there are outcroppings of coal, places where erosion or up-lifting has brought the edges of ancient sedimentary beds to the surface. Sometimes, loose chunks of coal accumulate next to these exposures, and at some point, someone must have thrown them into a campfire, only to discover that they would burn. The next step would have been to intentionally gather some of the loose coal for fuel, and then to start cutting chunks out of intact seams wherever they were exposed. Having begun as peat deposits laid down under water, many coal seams are more or less flat, and it is likely that early coal miners began by cutting short horizontal mine tunnels to follow the coal seams inward from their outcroppings. This kind of digging was relatively straightforward, although it could only extend a short distance before the roof of the tunnel began to collapse.

Where this kind of extraction was not possible, and where shallow coal seams could be found near the surface, it soon became common for miners to dig small open pit mines similar to water wells, a technique that had been originated in northwestern Europe as early as 4,000 BC by hunters seeking flints nodules to make stone tools.[3] Because they widened out as they went down, they were sometimes known as *bell-pits*. The soil and rock excavated to make them was usually piled on the ground nearby, and then the coal at the bottom of the pit was hauled up to the surface by human or animal muscle power. Once all the coal at the bottom of the pit had been dug out, short side tunnels were dug following the seam in every direction until all the coal that could safely be extracted had been taken out. At that point, another similar pit was dug a short distance away, a process which continued as long as there was coal that could be reached. The environmental impacts of this were significant, but fairly local in extent. The sinking of a pit destroyed all the vegetation on that spot, and the piling of the excavated materials smothered more vegetation nearby. The pits also altered drainage in the area, tending to lower the water table, thus making it hard for nearby surviving plants to absorb the same amount of moisture as in the past. In areas where large numbers of pits were dug, the whole landscape was also transformed, disturbing the movements of local wildlife.

Before industrialization, mining was mostly carried out by hand labor assisted only by draft animals, although where large amounts of water were available, hydraulic mining had also been employed since Roman times. By the nineteenth century, however, advances in steam engines and in metal-working had led to the creation of the first *steam shovels*, large machines featuring a movable digging bucket on the end of a long boom mounted on a heavy rotating base. This could be placed on a rail car for railway construction or on steel crawler treads for use on the ground. Steam shovels were able to move massive amounts of earth, and were, for example, the key technology employed in the construction of the Panama Canal between 1904 and 1914, one of the largest earth-moving projects in history.[4]

As steam shovels became more widely used in the early twentieth century, mineral hunters came to see that they offered the potential to create a new type of mining. Where deposits of ore or coal were more or less horizontal and not too deep below the surface, steam shovels could now strip away not only the overlying vegetation, but all the layers of soil and rock that lay beneath them. In the past, this had long been done by hand on a small scale for highly localized ore deposits, but now large plots of land could rapidly be excavated. For the immediate natural community this generally meant total destruction—not only were all the plants and animals removed, but also all of the soil, seeds and spores that might have allowed the ecosystem to regenerate. For the miners, though, operations became cheaper and safer, with no underground pits or tunnels to construct and no cave-ins or underground explosions or fires. It marked the beginning of large-scale *strip mining*.

Steam shovel mining grew rapidly in the early decades of the twentieth century, even as the steam engines powering them came to be replaced in the 1930s by more efficient diesels.[5] Individual machines grew larger and larger, until some were as tall as a 15-story building with the ability to excavate 150 tons of material in a single bite. Such machines allowed surface mining on a heretofore unheard of scale, and this, of course, allowed them to very quickly denude the land above the ore.[6] They worked in coal mines, but also in open pit copper and iron mines. In Bingham County, Utah, for example, an open-pit copper mine had been begun in the nineteenth century; it is still in use and currently penetrates almost a kilometer deep (over 3,000 feet) into the earth, while sprawling over a roughly circular area four kilometers across.[7] Developing such a mine disrupts the environment over a wide area. The non-ore material removed in the excavation has to be placed somewhere, burying the site's original natural communities. Then large amounts of refining waste will also be piled up, smothering other natural areas. Often these are not just physically massive, but contaminated with toxic heavy metals and processing chemicals which can leak into local watersheds.

In addition to steam shovels, dynamite and other explosives also came to be used in surface mining, just as they had in railway construction. In the Appalachian region of the United States, where extensive coal deposits were located, many at high elevation, a combination of explosives and mechanical excavators started being used in the 1960s to carry out a particularly destructive type of surface mining called *mountain top removal* (MTR). Typically, a coal seam would be located within a few hundred feet of the top of a hill and instead of expensively tunneling in from the side to extract the coal, the vegetation would first be cleared, after which explosives would be used to disintegrate tens or hundreds of feet of mountain top above the coal. The rock and soil that had overlain the coal bed would then be pushed to the side so that the coal could be excavated. Once this was done, the broken up material would either be piled back up on the hill top or else dumped into adjacent valleys.[8] Widely used in a number of states in the region, this practice caused a myriad of environmental problems from the destruction of vegetation and the removal of soil to the burying of stream headwaters and pollution of surface waters. The visual effects on the landscape in some areas were also devastating.

Shallow pit and strip mining for coal was feasible wherever seams lay horizontally near the surface. In many areas, however, all such deposits were soon exhausted, and long tunnels had to be dug either into hillsides or outward from the bottom of a deep pit. This was technically more difficult and more dangerous; in longer mine shafts or tunnels, it became harder to provide safe air to breathe, to pump out the groundwater constantly trickling in, and to avoid explosions of combustible gases released from the coal. Conditions far underground were much more dangerous for miners, although deep tunnel mining was less immediately destructive to the environment than was pock-marking the land with numerous shallow pits or stripping the land to get at the deposits. Regardless of how the shafts were excavated to get at the coal, it was rarely possible to recover all of it, and it was not uncommon for a third or more to remain in place either in the form of pillars and horizontal layers supporting the tunnel roofs, or in chronically flooded areas that could never be adequately drained. As long as active mining continued, any water seeping into the mine had to be pumped out as fast as it flooded in. But once the recoverable coal had been removed, pumping operations ceased, and the pits and shafts gradually accumulated water containing considerable amounts of dissolved oxygen. When this happened, new chemical reactions began to occur allowing mineral compounds present in the coal to oxidize and form water-soluble compounds. As long as the coal had lain in undisturbed beds sealed away underground there was little or no free oxygen present, and even the presence of water could not produce these reactions. But once oxygen was introduced, these reactions could begin and then continue for a long time.[9]

One striking example of such oxidation reactions can be seen with the mineral pyrite—iron sulfide. It is a common impurity in coal and is chemically stable as long as it remains completely buried. But once exposed to air, it can react with oxygen through several steps, partly triggered by the *feeding* on metallic ions of species of bacteria and archaea adapted to extreme environmental conditions similar to those that are presumed to have been present on the ancient earth. When this happens, the combination of chemical oxidation and microbial transformation of the iron in the pyrite lead to the production of sulfuric acid and a type of waterborne rust (iron hydroxide). As these build up, harshly acidic water carrying iron and other metals begins to flow out into the water table. Eventually it enters surface streams, killing fish and most other aquatic life and leaving behind tell-tale yellow to red iron precipitates known as *yellow boy*. This destructive process, which does not normally occur with undisturbed coal beds, was originally termed *acid mine drainage*. But since it can also be produced by piles of mining debris up on the surface, it is now known by the broader name of *acid rock drainage*.[10] Perhaps the most notorious example of it is the Rio Tinto (red river) in southwestern Spain, whose watershed has been the site of metal mining for 5,000 years. Not only is the river stained all over by iron deposits, leading to its name, but it contains water as acidic as lemon juice (pH 2), and is also heavily contaminated with a *witches' brew* of toxic metals.[11] (For comparison, it is worth remembering that most of the earth's water—that present in the oceans—is slightly alkaline, and that in comparison, the waters of the River Tinto are roughly one hundred thousand times more acidic.)[12]

Surface mining damages or totally eliminates existing natural communities at a site; underground mining often produces toxic drainage from abandoned mine tunnels and shafts that can destroy creatures living in and near streams, springs and rivers. Less commonly, un-dug coal in abandoned tunnels catches fire, leading to underground *coal seam fires*. Sometimes these can quickly be put out. Often, however, they become impossible to control due to one or more factors: flooded tunnels and the presence of poisonous gases exclude firefighters; it is difficult and expensive to completely seal abandoned tunnels; smoldering fires are able to follow small and irregular coal seams in many different directions for decades or even longer. Before active mining, such underground fires were very rare. But for a number of reasons they became much more common once coal began to be exploited on a larger scale. First, they were often accidentally started by the mining operations themselves, especially by accidental gas explosions. Second, economic realities led over time to the abandonment of thousands of partially air-filled tunnels containing unrecoverable coal, many with left-over piles of finely crushed material and coal dust piled near their entrance.[13]

Under these conditions, it became much easier for coal seam fires to start through accidents, lightning or spontaneous combustion. Once started, such fires have many different environmental impacts, ranging from the killing of surface vegetation and the animals that depend on it, to long-term releases of toxic fumes and soot, to widespread earth cave-ins due to overlying rock layers collapsing to fill the voids left by the burned away coal. Thousands of these fires are now burning worldwide, consuming many millions of tons of coal each year, and contributing by some estimates fully one percent of all the greenhouse gas emissions responsible for global warming.[14]

So far we have concentrated on the environmental impacts of coal production before it is even burned. Once mined, it almost always has to be transported some distance. In some cases, as with lignite going to nearby power plants to generate electricity, this may be a short trip—some power plants have intentionally been built very close to major coal deposits. In other situations, as with Australian coal being exported to China, it may travel thousands of miles. Coal is a heavy, bulky material, and is generally transported by barge or railway within countries, and by ocean-going ships in international trade. Moving it around, of course, takes energy and also generates some air pollution. The final destination for most coal today is a power plant or a cement kiln, and in either case, the material is usually ground into a fine powder and blown into a furnace along with lots of air. At the present time, coal-fired power plants are estimated to produce more than 40 percent of all the electricity used worldwide.[15]

Coal Burning

The burning of coal produces a wide variety of gaseous and solid waste products. The exact mix is determined by the type of coal, the firing temperature, the design of the furnace, the nature of any non-coal materials added to the fire, and the amount of air introduced. Most of the heat generated derives ultimately from the cellulose and lignin that were the main components of the plants that made up the original peat. Since both of these contain major amounts of carbon and hydrogen, their complete burning produces large amounts of carbon dioxide and water vapor. Green plants also contain smaller amounts of a wide range of materials including proteins, waxes, essential oils, pigments and incorporated mineral compounds. The remains of these substances add other chemical elements to coal, particularly calcium, potassium, phosphorous, magnesium, nitrogen and sulfur. In addition, plants grow in soil, and even the wet swampy soils of coal forests were rich in elements like silicon, iron, and aluminum. Finally, long-term burial also often led to the presence of radioactive atoms like barium, uranium and thorium in coal along with some toxic lead, arsenic, mercury and cobalt. (A small fraction of potassium atoms are also radioactive.)[16]

The ash produced by burning coal is generally divided into larger, heavier particles which collect at the bottom of the furnace, and smaller or lighter particles which move upwards in the flow of exhaust gases. The former is a combination of fine bottom ash and coarser material called *slag* or *clinker*. The lighter, airborne emissions are known as *fly ash* and *soot*. Bottom ash usually contains a number of toxic substances, but since it is a solid, it does not normally escape from the coal furnace, and some uses have been found for it in the manufacture of cement, asphalt and other products. If disposed of properly, it poses low levels of risk, although all too often it is stored uncovered out of doors or in slurry ponds, where it can contaminate nearby areas. In February 2014, for example, in the American state of North Carolina, failures of maintenance, monitoring and regulation led to the release of at least 39,000 tons of coal ash and polluted water from a failed retention pond adjacent to a closed coal-fueled power plant. The release contaminated more than 70 miles of the Dan River in North Carolina and adjacent Virginia with arsenic and other toxic substances, causing significant environmental damage and endangering the drinking water supply for several communities.[17]

The lighter wastes of coal burning that are present in the exhaust present a different set of risks. Fly ash includes enormous numbers of microscopic particles, many of which are so tiny that they can pass deeply into human or animal lungs. Most consist of oxides of silicon, aluminum and calcium. But along with these come particles containing toxic metals like arsenic, lead, cobalt, mercury, radioactive uranium and thorium and small traces of harmful dioxins, although it is not fully clear how much harm these cause.[18] In addition, though the mechanisms are not completely understood, there is also evidence that inhalation of small combustion particles contributes to the development of both heart and lung disease.[19]

Soot—tiny particles of carbon with unburned aromatic compounds stuck to them—is another harmful exhaust product. Along with fly ash, it used to routinely came shooting out of chimneys in homes, factories and railroad locomotives, darkening the environment all around and killing sensitive plants and animals. In 1775, soot was also the first substance to be clearly identified as a cause of cancer. In that year, the British surgeon Percivall Pott showed that young British chimney sweeps, who were constantly exposed to soot in chimneys, developed cancer of the scrotum at rates dramatically higher than boys working in other trades.[20]

Besides ash and soot, the burning of coal releases large amounts of invisible gases, and the unraveling of their impacts offers a good illustration of how our understanding of the workings of the natural environment has grown over time. As early as the mid-nineteenth century, early progress in chemistry made it possible for a small number of working chemists to tackle a number of practical problems. In Britain, one such individual was Robert Angus Smith, who was based in the rapidly expanding textile manufacturing center of Manchester. As early as 1852, Smith demonstrated that severe air pollution was resulting from the exhaust gases coming out of coal-fired steam engine boilers and furnaces. In that era, and indeed up until the mid-twentieth century, Manchester was notorious for smoke, soot, dirt and chronically gray skies. Architectural stonework and outdoor statuary in the city tended to be eaten away, and many people experienced frequent bronchitis and other respiratory illnesses. Natural vegetation and wildlife in the region was also heavily damaged.[21]

As we saw in Chapter 4, in 1863 the British Parliament passed the Alkali Act, one of the first modern air pollution control laws. It was designed to limit the hydrochloric acid emissions from Leblanc process chemical plants. After it was passed, Robert Angus Smith was appointed to head up a small office set up to oversee the act's monitoring and enforcement.[22] In 1872, Smith's book *Air and Rain: the Beginnings of a Chemical Climatology* helped pioneer the broader study of air pollution, and for the first time used the term *acid rain*.

Despite Smith's pioneering research and the passage of the Alkali Act, for many decades little was done in Britain to reduce the harmful effects of coal emissions. This only began to change in the early decades of the twentieth century when the spread of grid electricity meant that larger and larger, coal-burning power plants started to be built. Demand for electricity was particular high in cities, and so the first serious attempt to control the sulfur dioxide air pollution resulting from coal burning was undertaken in London. It took the form of a kind of *scrubber* for removing some of the sulfur compounds from the power plant exhaust before it was released into the atmosphere. Known as a *flue gas de-sulfurization* unit, the first important one was installed in a London power plant in 1931. During and immediately after World War II, economic restrictions led to the technology being dropped—a step backward that ended up having serious consequences for public health. In the 1970s, however, it was revived and then fairly widely adopted in rich countries around the world. In its most common form, it runs the exhaust fumes through a wet bath containing powdered limestone or lime. Under these conditions, some of the sulfur dioxide combines with the lime to form gypsum (calcium sulfate), which is insoluble and also has some commercial use in the making of plaster board. By some estimates, such scrubbers are able to remove up to 95 percent of the sulfur from the coal, with the rest ending up in the atmosphere.[23]

Flue-gas desulfurization is a proven technology, but it requires additional maintenance, supplies and the handling of the sulfates produced by the scrubber. This adds to the cost of operating the plant, and increases the price of the electricity it produces. In this way, economics usually end up determining how much clean-up will go along with the large-scale burning of coal. Most, if not all, of the numerous pollutants produced by coal fires can be captured by a wide range of chemical scrubbers. But the cost is so great that eliminating most of them makes the power produced prohibitively expensive.

Serious air pollution resulting from the burning of brown or bituminous coal is not a new problem. As long ago as 1285 the English king created a commission to investigate London's poor air quality, and some efforts were made over the following decades to limit coal burning in order to reduce pollution. Enforcement, however, was lax, and over the subsequent centuries, as deforestation made firewood and charcoal less and less available, increasing amounts of coal continued to be burned both in London and throughout the rest of England.[24] In 1661, for example, the famous diarist and early environmentalist John Evelyn published a pamphlet titled *Fumifugium* which was intended to drawn attention to the extreme dirtiness of London's air and to various steps that could be taken to improve it.[25] Though much discussed, *Fumifugium* does not seem to have led to many practical changes. Instead, deadly smogs continued to be recorded in that city right up to the middle of the twentieth century, with especially bad episodes occurring in 1813, 1873, 1880, 1882, 1891, 1892 and 1948.[26]

Limited new smoke regulations were enacted in Britain between 1845 and 1926, but these were not extensive enough to eliminate the repeated episodes of harmful pollution in London and other British cities. Finally, in 1952, in an era when Britain was still recovering from the devastation and heavy cost of World War II, the crisis of air pollution came dramatically to a head. On Friday December 5, visibility in London began to decline and soon fell to near zero. The city became more or less paralyzed as conditions developed which later became known as the Great Smog. Travel became impossible and most normal activities were halted or curtailed for a period of five days—several people even died by drowning in the Thames; they fell in because they could not see that they were on the riverbank. Besides smoke, soot and water vapor, the atmosphere became heavily loaded with sulfuric acid from the sulfur in the coal. Economic losses were significant, but the effects on public health were even more devastating, with early counts finding that 100,000 people had been sickened and 4,000 had died. Later research in fact suggested that the number of deaths caused by the Great Smog may actually have reached 12,000.[27]

By the mid-1950s in Britain, the impact of the Great Smog, combined with advancing scientific understanding and slowly improving economic conditions led to a new focus on the state of the environment, and by mid-decade this translated into the Clean Air Act of 1956, the country's first comprehensive air pollution control legislation.[28] That bill contained a number of provisions which reduced the emission of dark smoke, soot and sulfur, especially in urban areas, while also mandating taller chimneys for large facilities that burned coal. Taken together, these reduced urban air pollution to the point where there has never been a repeat in Britain of the deadly Great Smog of 1952. Other countries, however, have not been so fortunate, and in a number of developing nations, especially in Asia, deadly smogs have continued to occur right up to the present day due mostly to the burning of coal and to the (often illegal) clearing of tropical forests by intentionally set fires.

Britain's Clean Air Act of 1956 was a forerunner of other significant air pollution regulation to be passed later in the twentieth century. During the nineteenth and twentieth centuries, the United States also experienced significant air pollution from coal burning with several locations including Pittsburgh, Pennsylvania becoming notorious for poor air quality. For the most part, however, mostly due to geography and luck, the country did not experience the particularly deadly smogs that had long plagued London and other British cities. One notable exception to this, however, occurred in the industrial town of Donora, Pennsylvania, about 30 miles south of Pittsburgh, where coal was widely burned and where a zinc works and a steel wire plant released industrial emissions including hydrogen fluoride and sulfur dioxide. Air quality in Donora had long been a problem, but in late October of 1948 a temperature inversion trapped the pollution close to the ground, and over the next three days, nearly half of the town's 14,000 people became sick, with twenty people and hundreds of animals dying outright.[29]

Fourteen years later, in 1962, Rachel Carson's pioneering book *Silent Spring* became a best-seller. After that, the state of the environment had become one that American politicians could no longer ignore. In response, in 1963 the United States passed its own *Clean Air Act*, although initially it called primarily for studies of air pollution rather than for regulation of emissions. By 1970, the act had been expanded to include both the determination of appropriate standards for different air pollutants and the enforcement of those levels through a new national office known as the United States Environmental Protection Agency (USEPA). It was further expanded again in 1977 and 1990 in response to the continuing growth in scientific understanding of air pollution.[30]

The USEPA played a key role in enforcing reductions in harmful ozone-destroying refrigerants (see Chapter 4), and also has responsibility for regulating excessive noise in urban areas. Important categories of air pollutants that it helps to regulate include soot and particulates, ozone, lead, sulfur and nitrogen oxides, and the fumes from gasoline and from other volatile organic materials that are commonly used in paints. It is responsible for emissions from both stationary and mobile sources of pollution which means that it regulates not only what comes out of smokestacks, but also out of cars, trucks, ships and airplanes. Currently it is seeking authority to regulate carbon dioxide emissions, although that remains politically controversial.[31] In addition, the USEPA has nationwide responsibility for the prevention of water pollution in both ground and surface waters, deriving its authority from 1972 legislation known as the Clean Water Act. Overall, it has been extremely successful in reducing pollution in the United States, although this has necessitated the creation of a large bureaucracy with roughly 15,000 employees and a budget which stood at close to eight billion dollars in 2015.[32]

Acid Rain

Sulfur dioxide emitted by burning coal was an important hazard in Robert Angus Smith's Manchester and also played a key role in London's deadly Great Smog of 1952. Since pre-cleaning the coal or capturing the harmful emissions was expensive, one tactic taken by some facilities burning large amounts of it was to build extremely tall smokestacks in order to spread the pollutants over a wide area. This approach had been pioneered by metal smelters starting in the nineteenth century in order to disperse toxic substances like arsenic and lead that were liberated when metal ores were smelted (see Chapter 4). From a societal point of view, tall smoke stacks were a limited success in that they reduced the worst pollution in the immediate area around the smelter or power plant. From an environmental standpoint, though, they were largely a failure, because they did not capture or remove pollutants, but rather spread the harm over a wider area, often at considerable distance from the plant. In the United States for example, at the end of 2010, there were 284 smoke stacks taller than 500 feet (152 meters) in height located at 172 coal-fired power plants, with more than a dozen exceeding 1,000 feet (305 meters).[33] In the US, many were located not far from the Appalachian Mountains where extensive beds of coal were located.

By the 1970s, some of the damaging impacts of long-distance transport of pollutants like sulfur dioxide and nitrogen oxides from coal-burning plants had begun to be documented. Some of the earliest observations were of the disappearance of trout and other fish from high-altitude lakes in the Adirondack Mountains of New York, and of dying spruce forests in the Green Mountains of Vermont. Eventually, intense research programs carried out by foresters, atmospheric scientists, lake biologists and soil scientists succeeded in demonstrating that one important cause of this damage was the emissions from large coal-burning power plants upwind of the affected areas. In spite of the distances from the sources, rain and snow turned acid by dissolved pollutants was coming down on these extensive natural areas, harming animals and plants, and changing the chemistry of water and soils. (A similar phenomenon was also documented in central Europe, where the German term *Waldsterben* or *forest death* was used to describe it.) Sulfur dioxide was the primary culprit, although nitrogen oxide emissions also played a role. The former could combine with water to make highly acidic, corrosive sulfuric acid; the latter made aggressive nitric acid and also sped up the process by which sulfur dioxide was converted to sulfuric acid.[34] Over the following decades, the revised Clean Air Act of 1990 stipulated reductions in the amount of pollutants that could be released into the atmosphere from coal burning, and this has contributed to gradual improvement in overall air quality in eastern North America. Red spruce (*Picea rubens*), for example, one of the tree species most heavily impacted by air pollution in the past, is once again growing well.[35]

Petroleum Extraction, Transport and Refining

In trying to understand the potential environmental problems associated with the extraction and use of petroleum, it is helpful to take a closer look at its origin in the earth's crust. In Chapter 4 we described how petroleum forms from organic matter such as algae and zooplankton that accumulates in sea floor or lake bottom sediments. One key to its formation is the failure of some of the organic matter that rained down to fully decompose under the conditions present on the bottom.

As with plant peat destined to eventually become coal in freshwater swamps, organic matter on the seabed or lake bottom comes to be buried over time by overlying mineral sediments whose mass eventually presses down very hard on the partially decomposed remains. Over long periods of time, through pressure and the heat present as layers became buried more and more deeply, the animal and plant proteins, fats and carbohydrates became chemically transformed into crude oil, a complex mixture of materials which also contains a number of toxic substances.

Drilling for petroleum involves penetrating into layers of rock containing petroleum which is usually under great pressure. Often the oil lies adjacent to a *salt dome*, an unusual geological formation that forms when columns of saline deposits flow slowly upwards from deeply buried salt-rich layers underneath, typically under an existing or former ocean bed. For these reasons, new wells which are drilled down to levels where petroleum is present often encounter oil that is under great pressure and is frequently contaminated with salt water. The pressure can cause *blow-outs*, or forceful explosions of hazardous petroleum from the top of the well; the salt can also be a cause of groundwater contamination wherever oil wells are drilled on land. Because they extend down through many layers of rock, oil wells may also encounter a variety of other hazardous materials like hydrogen sulfide and heavy metals, and these too are potential sources of serious pollution. Over time, as the original high pressure in an oil well falls, continued production may require massive pumping of water and other fluids into the well in order to maintain adequate pressure. Efforts are generally made to re-circulate these fluids, but some leakage often occurs with more or less local contamination by oil, salt water or both.[36]

After the oil has emerged under natural pressure or been pumped out, it must be transported, stored, refined and blended into products. The oil industry's overall record is mixed. On the one hand, thousands of oil wells have been dug and brought into production without too much environmental damage, and oil production and utilization have contributed to considerable economic expansion throughout the twentieth century. It is also hard to deny that many of the conveniences of modern life have been built on a platform of refined petroleum products. At the same time, the history of oil has been repeatedly marked by episodes of significant harm to the environment and to human health (see also Chapter 4).

Outside of the creation of the world's worst oil release when the Iraqi military intentionally blew up hundreds of Kuwaiti oil wells during the Gulf War in 1991,[37] the most dramatic incidents of environmental damage have involved unintentional spills of crude oil or refined petroleum products. Typically, these have occurred under two circumstances: when newly drilled but inadequately controlled oil wells reached underground petroleum reservoirs that were under great pressure, sending oil hurtling to the surface; and when shiploads or large tanks of already extracted oil broke open. In 1901 and 1902, for example, there were notable spills on Spindletop Hill near Beaumont, Texas in the US, the first occurring in January 1901 during the initial drilling of a well into a salt dome. At 1,139 feet (347 meters) of depth, oil under high pressure was encountered, forcefully blowing the drilling pipe out of the well and producing a geyser of oil that rose more than 100 feet into the air. Oil flowed out at an estimated rate of 4,200,000 gallons per day, and it took more than a week before the well was completely capped. Needless to say, considerable local contamination resulted from this incident. In September 1902, another *gusher* at Spindletop was accidentally set on fire by an oil worker, and the resulting blaze burned for roughly a week creating extensive air and soil pollution.[38]

Spindletop turned out to lie on top of one of the largest oil fields in North America, and during the twentieth century it produced vast amounts of petroleum—by 1985, more than 150 million barrels had been pumped out of it.[39] In response to its output and to that of other oil fields in the region, the oil refining and petrochemical industries built a long string of refineries and chemical plants between Houston, Texas, and New Orleans, Louisiana. Eventually, this petrochemical cluster developed into the largest concentration of oil-based industries in the United States, and one of the largest in the world.[40] This area, which previously had been clean pine forests, wetlands, pasture and crop fields, experienced significant pollution over time, and became the setting for long-running battles involving environmentalists, advocates for the poor, state and federal regulators, and the petrochemical industry.[41] Complaints included charges of groundwater contamination, the presence of abnormally high rates of asthma due to air pollution and extra cases of childhood leukemia resulting from elevated concentrations of benzene and 1,3-butadiene in the air. A number of public health studies have been carried out in the area, mostly without clear-cut results. It is, however, known that air pollution from petrochemical plants releasing volatile organic compounds and particulate matter can cause elevated rates of asthma and other breathing problems in children.[42]

In recent decades, there have been more spills from damaged oil tankers than from oil well blow-outs. Mostly this is because the mechanical and hydraulic technologies needed to control oil under great pressure at the well-head have dramatically improved since the days of Spindletop. In the earliest days of the oil industry, petroleum was transported in the same way that whale oil had been—in wooden casks. As volumes grew, however, this proved to be a cumbersome and unworkable system, and by the late nineteenth century, specialized oil tanker ships had been developed. These contained multiple interior metal compartments that could be filled with oil and then pumped out upon delivery. With many techno-logical refinements, this remains the essential design of oil tankers today, some of which are among the largest ships ever built, with the biggest now able to carry as much as three million barrels (126 million gallons) of oil. The largest are so big that they are unable to enter most harbors, and instead must off-load their cargoes at specialized, deepwater oil terminals. These behemoths are used to haul vast cargoes of crude oil many thousands of miles across the oceans, and they are supplemented by an even larger number of smaller petroleum tankers and barges plying coastal and inland river routes, some carrying crude oil, but most carrying refined products like gasoline, kerosene or diesel.[43] Spills can occur from any kind of oil transport vessel, and every year some of these occur all around the world due to bad weather, equipment failure or human error.

Worldwide, the petroleum industry expanded enormously during the second half of the twentieth century, much of its growth fueled by the tremendous increase in the number of motor vehicles and civilian aircraft. This, in turn, led to dramatically increased shipment of oil in larger and larger ships. Many efforts were made to increase safety and reduce spills, but in spite of this, quite a few episodes of significant environmental damage still occurred. In 1979, for example, two large oil tankers, the *Aegean Captain* and the *Atlantic Empress* collided in the fog near the island of Tobago in the Caribbean. Both ships were seriously damaged and both caught fire, but the crew of the *Aegean Captain* managed to put out the fire and stabilize their ship. On the *Atlantic Empress*, however, many of the crew were killed and fires raged out of control. The ship burned and leaked oil for more than two weeks and then finally sank releasing the rest of her nearly 287,000 tonnes (approximately 90 million gallons) into the sea, the largest oil tanker spill to date.[44]

Not much scientific research was done to document the impact of the enormous spill from the *Empress*, but a much smaller spill that occurred ten years later in Alaska was closely monitored almost from the start. On March 24, 1989, the medium-sized tanker *Exxon Valdez* ran aground in Prince William Sound spilling approximately 35,000 metric tonnes (roughly eleven million gallons) of crude oil into the rich, cold, coastal waters of the North Pacific. Because of the many coves and bays in that area of Alaska, the oil ended up coating roughly 1,300 miles of coastline along with more than 10,000 square miles of ocean. The entire ecosystem, from bacteria up to killer whales was heavily damaged and losses included the deaths of more than 100,000 seabirds and roughly 3,000 sea otters, along with very large numbers of salmon and smaller numbers of other marine mammals and birds. Even after extensive clean-up efforts costing more than a billion dollars and the passage of 18 years, studies in the 2000s suggest that at least 26,000 gallons of oil still remained as contaminants along the affected coastline.[45]

In the wake of the *Exxon Valdez* disaster, the United States passed legislation designed to avoid similar catastrophes in the future. One important component of this was the requirement that the single-hulled tanker fleet gradually be replaced with double-hulled vessels that would be more resistant to leaking. This structural standard also came to be formalized in connection with an earlier (1983) international agreement accepted by more than 150 nations, called the *International Convention for the Prevention of Pollution from Ships*. In what is clearly an important step towards better environmental protection, parties to the agreement have agreed to phase out all single-hulled oil tankers by 2026.[46]

The Spindletop *blow-out* occurred in 1901 in the early days of the modern petroleum industry. But even after more than a century of engineering advances and many significant improvements in safety, catastrophic petroleum spills causing great environmental damage still seem to be an unavoidable part of drilling for oil. One reason for this involves the depletion of many oil fields on the continents and the re-focusing of exploration on oil deposits located deep underneath the sea floor. In most cases this involves the use of large, floating oil well drilling platforms capable of first lowering equipment through thousands of feet of ocean before they begin to drill additional thousands of feet down into the earth's crust.

In 2010, one such platform, known as the Deepwater Horizon, was at work in the Gulf of Mexico about 40 miles off the coast of Louisiana, under contract to the large petroleum company BP. On April twentieth its drill bit was down nearly 7 miles below sea level when it encountered natural gas—methane—under high pressure. (Some methane is almost always present in oil deposits.) As a result of deficiencies in both the quality of materials and workmanship used in the construction of the well, and in the functioning of safety devices, the methane escaped up the well onto the drilling platform where a spark caused it to explode into a fireball that killed eleven workers and destroyed the drill rig. In addition, a device located on the sea floor designed to shut off the well in the case of such an accident did not operate properly, allowing oil under high pressure to come squirting up into the ocean. Because the well head lay more than 5,000 feet underwater, it was impossible for divers to make repairs, and this gigantic spill went on for nearly three months (87 days) until a huge, specially built well cap could be lowered into place to finally stanch the flow. Over that twelve and a half week period, 780,000 metric tonnes (about 210 million gallons) of crude oil entered the ocean ecosystem of the Gulf of Mexico, shutting down fisheries in a large part of the Gulf, fouling beaches with oil, and causing billions of dollars in economic losses in coastal settlements in four American states. (The total release from the Deepwater Horizon blow-out was more than twice as great as that from the previous record spill from the *Atlantic Empress* in 1979.) Studies on the damage to human health and to living communities are still underway, so completely final conclusions cannot yet be drawn. But early evidence suggests widespread death and disease among sea creatures occurring even months after the well was capped, along with significant health impacts both for clean-up workers and for children and adults around the Gulf Coast. In April 2016, a court considering the many lawsuits against the oil company concluded that adequate environmental restoration was likely to cost between eight and nine billion dollars.[47]

Carbon Dioxide and the Atmosphere

During most of the nineteenth and early twentieth centuries, smoke, soot and sulfur were the most obvious problems caused by coal fires, and except for a few scientists, no one seems to have thought much about the impacts of coal burning on earth's overall environment. Yet even in the 1890s, the worldwide impacts that massive carbon dioxide emissions from fossil fuel use might have on global temperature were already foreseen by some insightful researchers.

Starting in the 1700s, a number of observers in Switzerland had suggested that the unusual distribution of boulders in several high valleys at the foot of alpine glaciers was the result of glacial transport, followed by the receding of the ice. After further study, this idea came to be accepted, and by the mid-nineteenth century, scientists had begun to speculate that there had been *ice ages* in the past, periods of time when glaciers covered far more land than they did in the 1800s. Research on glaciers was then begun in several different parts of the world, and by the 1870s many scientists had come to agree that widespread epochs of continental icefields had in fact occurred.[48] Having reached this conclusion, various scientists began to propose possible mechanisms that could have produced such a general global cooldown, with some people also raising concerns that renewed glaciation in the future might undermine human progress.

Quite a few researchers were involved in this work, including the Swedish chemist and physicist Svante Arrhenius, who had also made notable contributions to our understanding of the underlying nature of ions in solution and of acids and bases. In 1896, building on the earlier work of the French mathematician and physicist Joseph Fourier, and incorporating recent meteorological research from the United States, Arrhenius proposed that the earth's temperature would rise in the future as the concentration of carbon dioxide in the atmosphere increased. His explanation for this was that even though carbon dioxide was transparent to visible light, it was still capable of reflecting downward some of the infra-red radiation (recycled solar heat) re-radiating from the earth's surface. When it did this, the energy it sent back downward raised the atmosphere's temperature, so that the invisible carbon dioxide would behave a bit like the glass roof of a greenhouse. And since human burning of both fossil fuels and firewood had been rapidly increasing in the nineteenth century, Arrhenius argued that it was highly unlikely that a new ice age would come about any time soon.[49] More than a century of research since then has shown that Arrhenius was basically right, and it seems remarkable that as long ago as 1896 the fundamental relationship between carbon dioxide emissions and global warming had already been worked out and made public.

The nineteenth century came to an end just four years after the publication of Arrhenius's work, and the first six decades of the twentieth century were marked by the widespread adoption—at least in rich countries—of dramatic new technologies like electric lights, telephones, automobiles and assembly lines. They also witnessed the unprecedented human and economic losses of the 1918 influenza pandemic and the two world wars. (An estimated fifteen million were killed in World War I, while World War II is believed to have resulted in at least 50 million deaths. In addition, roughly fifty million are thought to have died in 1918 during the global outbreak of Spanish influenza.) Despite these appalling losses, overall technological progress continued, and in the case of war-related work, even accelerated. Particularly in the years following World War II, mass immunization against infectious disease, provision of safer water supplies and increased agricultural productivity all led to increased human numbers, with global population doubling between 1880 and 1960 from one and a half to three billion (see Chapter 3). At the same time, overall economic activity grew even more rapidly, leading to enormous expansion in the burning of both coal and petroleum-based fuels like gasoline and diesel.

Detecting Global Warming

It may seem hard to believe, but until quite recently in human history no one could ever be sure exactly how hot or cold anything was—there was simply no way to accurately measure heat, and this presented a significant barrier to the advancement not only of science, but of technology in general. Today, of course, we take the existence of accurate and inexpensive thermometers for granted. The earliest temperature measuring devices, however, date back only to the latter part of the sixteenth century, when scientists like Galileo started to develop them to assist in their experiments. After that, it still took more than a century before the German physicist and glass blower Daniel Gabriel Fahrenheit finally succeeded around 1710 in building the first practical glass-tube thermometer marked with numbered lines. And even after Fahrenheit, many decades more were to pass before modern temperature scales (Fahrenheit, Celsius, Kelvin) were standardized and before enough skilled craftsmen had been trained to produce a regular supply of thermometers. Only in the middle of the nineteenth century did accurate instruments became readily available, a development that gave a tremendous boost to scientific research in general and to meteorology (weather science) in particular.[50]

From the seventeenth century on, the far-flung British Empire had been tied together by large number of merchant sailing ships protected by a smaller number of warships. Both fleets faced constant hazards due to storms, yet as late as 1800 sailors really had no good way to predict the weather. By the middle of the nineteenth century, however, advances in instruments had begun to change this, and in 1854 Britain established a *Meteorological Office*, the first modern weather forecasting bureau, whose primary goal was to predict coastal storms which posed a danger to shipping.[51] In the following decades, several other nations followed this lead, with the United States creating its *National Weather Service* in 1870 with a mission of predicting both coastal and inland weather.[52] Agencies like these opened networks of stations in different locations and routinely recorded data such as temperature, air pressure and wind speed, with the result that even before the end of the nineteenth century, an accurate record of earth's air temperature began to be kept in many places around the world.

The main goal of this meteorological work was to improve weather predictions in order to benefit seafarers, farmers, and others whose work and livelihood was impacted by storms and other weather events. But in the twentieth century, once serious scientific study of climate patterns had begun, these long-running baseline records also proved essential to working out the ways in which earth's climate had been changing in response to human activities. By the middle of the twentieth century, for example, there was a roughly one hundred year long record of accurate, instrument-based temperatures from many locations that clearly showed a statistical trend towards warming.[53] This was powerful evidence. Yet knowing that the earth was millions of centuries old, scientists were eager to obtain even longer temperature records dating back before formal weather records had begun to be kept. Clearly, there was no way to travel back in time with good thermometers, so instead scientists undertook a serious search for indirect measurements of past temperatures.

In the mid-1950s, a Danish geophysicist named Willi Dansgaard speculated that air bubbles frozen inside glacial ice might contain samples of the earth's atmosphere from different dates in the past. He further suggested that the makeup of the atmosphere before the keeping of records began could be determined by careful analysis of bubbles found in columns of ice removed from ancient glaciers like those in Greenland and Antarctica. Obtaining ice cores from such places, however, involved expensive, complex drilling rigs and logistical challenges far beyond what normal scientific research could manage. For roughly a decade, therefore, Dansgaard's thinking remained completely speculative. By the mid-1960s, however, an opportunity arose to obtain ice cores from Greenland that had been drilled by the US military for other purposes. Upon analysis, Dansgaard was able to show that the concentration of different gases in the atmosphere had in fact varied over the thousands of years covered by his samples. In particular, both the concentration of carbon dioxide and that of two different isotopes of oxygen had measurably changed over time. This was an important breakthrough, and formed the beginning of a new specialized field of scientific research operating at the interface of geophysics, atmospheric science and climatology.[54]

Oxygen exists naturally as three isotopes, common oxygen-16, and much rarer oxygen-17 and oxygen-18. (Isotopes are different physical forms of the same chemical element, differing only in mass but not in chemical properties.) Oxygen-18 weighs about 12.5 percent more than oxygen-16 and so it behaves differently in the ocean and in the atmosphere. When temperatures are warmer, slightly higher concentrations of oxygen-18 are found in the air; under cooler conditions, oxygen-18 is not quite as common. By calculating the ratio of oxygen-18 to oxygen-16, it became possible to track changes in atmospheric temperatures going back thousands of years. The longest such record found so far comes from ice cores drilled out of polar ice in East Antarctica where the snow pack has been accumulating for roughly 800,000 years.[55]

Around the same time that Dansgaard was planning his research, another atmospheric scientist was beginning a very different program of data gathering which also ended up shedding light on the changes earth's climate was experiencing under human impact. His name was Charles Keeling, and he was a chemist turned geochemist and oceanographer. In 1958, he started a systematic, open-ended program designed to directly measure the concentrations of the different gases making up earth's atmosphere. After careful planning, many of his measurements came to be made high up on the slopes of 13,679-foot Mount Mauna Loa in Hawaii, a site chosen for its remoteness from large artificial sources of gas emissions. The results of his observations constitute the longest and best record of its type anywhere on Earth, and they clearly show dramatic changes in our atmosphere, especially in the amount of carbon dioxide present. In 1958 when Keeling began, his measurements showed a concentration of atmospheric CO_2 of

315 units (parts per million). Forty-seven years later, at the time of his death in 2005, CO_2 had risen steadily to 380 units; by 2016 it had risen to 405 units—a rise of almost 30 percent since 1958—and was still rising.[56]

By the twenty-first century, there was strong agreement among most atmospheric scientists that global temperatures were rising, and that the release of carbon dioxide from burning fossil fuels and other human activities was a major cause of this increase. The accuracy and reliability of Keeling's data had been corroborated by similar gas sampling programs at other locations, and Dansgaard's results correlating atmospheric temperature with concentrations of CO_2 had proved to be in line with later research analyzing glacial ice core samples from Antarctica. Together, these two different ways of measuring atmospheric carbon dioxide were able to provide fairly accurate measurements dating back hundreds of thousands of years.[57] (Less precise techniques based on the makeup of rocks and fossils make it possible to chart the broad changes in atmospheric CO_2 over many millions of years. And these too have consistently shown a strong relationship between the amount of CO_2 present and earth's average temperature.)[58]

Before the industrial revolution of the late 1700s, average atmospheric CO_2 concentration had been roughly 285 units. Throughout the nineteenth and early twentieth century it had risen relatively slowly because at first much of the extra CO_2 produced tended to dissolve in the oceans rather than be stored in the atmosphere. Over the course of the twentieth century, however, the sea's capacity to further absorb carbon dioxide slowly began to decline, while atmospheric emissions of CO_2 and other greenhouse gases rose rapidly. With more CO_2, the atmosphere retained more heat, and global climate grew warmer. These changes were not completely smooth—there were intervals when warming paused or even reversed. But the long-term trends of more carbon dioxide and higher average temperatures seemed to follow inexorably from greater fossil fuel use and increased production of other greenhouse gases including methane and nitrous oxide.

The overall results of these changes has been well summarized by the USEPA: global temperatures are increasing; in response, earth's climate and weather have been changing, as have been conditions in the oceans and ice sheets. Human activities, especially the burning of fossil fuels, are the major cause of these changes.[59]

Effects of Global Warming

So how will a warming world be different? Summarizing the different impacts will not be easy because increasing average temperatures will produce a complex pattern of changes, many predictable, some merely probable, and others probably unanticipated. These will vary greatly from region to region, and most people will experience climate, and the weather patterns that contribute to it, only in their own home region, and one day at a time. In addition, many northern countries are located in historically cool to cold climates in the north temperate zones, where the idea of a warming world might at first seem to promise relief from harsh winters. The full effects of global warming will thus be multi-faceted, although several broad and destructive patterns have already become quite clear.

Glaciers and ice sheets are melting in most parts of the earth, with mountain glaciers receding uphill as they did in Switzerland at the end of the ice age, and earth's two polar ice sheets melting notably around their edges. In America's Glacier National Park, for example, there is documentary evidence that in 1850, roughly 150 sizeable glaciers were present, and that sixty years later when the park was established in 1910 nearly all of them were still there. But in the hundred years to 2010, more than 80 percent of them had disappeared, leaving only twenty-five remaining in the early twenty-first century. Even those are shrinking fast, and by the middle of the present century, *Glacier* National Park may no longer be an appropriate name for this spectacular place in the northern Rocky Mountains.[60]

Of earth's two great land ice sheets, the smaller is in Greenland where it covers an area nearly the size of Mexico with ice averaging two kilometers deep. Were it all to melt, it would release enough water to raise global sea levels more than 20 feet. This would flood large swathes of the world's coastal areas where more than a billion people live, and almost certainly cause the partial abandonment of many large, critical port cities such as London, Mumbai, Kolkata, Manilla, Hong Kong, Shanghai and New York. And beyond the cities that would be wholly or partly flooded, huge areas of valuable coastal wetlands—salt marshes, mangrove forests, etc.—that are critical for the survival of fisheries and wildlife populations would also be destroyed, including unique natural areas like Florida's Everglades and the Sundarbans of India and Bangladesh. Nor would these readily be replaced by coastal wetlands a bit farther inland, since most such areas are highly valuable agricultural land that would likely come to be protected by dikes rather than made available for conservation.[61]

In Antarctica there is a vastly larger ice sheet which covers 98 percent of that continent's land, an area roughly fifteen times as large as Greenland. Antarctica is significantly colder than the Arctic, and so far its ice is not melting as fast as Greenland's. But if all of it were to melt, it would raise ocean levels an estimated 200 feet, wiping out a much larger portion of earth's coastal zones including extensive farmlands, cities, etc. It is also important to point out that melting glaciers are not the only reason that sea levels have recently been rising at a rate of roughly one foot per century. Higher temperatures not only cause more ice to melt and run into the ocean—the melting of floating sea ice does not enter into this equation because its weight already raises sea level—it also causes seawater to expand in volume simply because it is warming up. A foot a century may not sound like much, but in the foreseeable future, these rises are likely to cause significant damage to human civilization.[62]

In addition to rising sea levels, many scientists believe that global warming is also influencing the frequency of heat waves, damaging droughts, wildfires and storms. As is always the case with the weather, these changes show up through a combination of shifting long-term averages and individual severe weather events. In 2003, for example, a long-lasting heat wave in France and surrounding countries produced the highest average summer temperatures for at least 400 years. This resulted in an estimated 20,000–70,000 human deaths, mostly among the elderly, with France being particularly hard hit.[63] Only three years later, 2006 once again saw all time high summer temperature records broken in several countries a bit further north in Europe; in 2010, widespread intense summer heat in a band right around the northern hemispheres caused significant damage in many countries, most notably in the Russian Federation. In Russia that year, beginning in spring with drought and warmer temperatures than had ever been recorded, the end of July saw the ignition of hundreds of wildfires in forests, drained peat bogs and other habitats, producing weeks of harmful air pollution. All told, an estimated 56,000 people died, while heat and drought led to widespread crop failures. That, plus direct fire damage, caused additional economic losses costing many billions of dollars.[64] In these events, and in other similar heat wave/droughts, there is understandable focus on loss of life, on economic costs including crop losses, and on human inconvenience. But wildfires and droughts don't just kill people; they also kill millions of wild animals and plants, creatures that play essential roles in the continuation of essential natural cycles on earth. In addition, many experts believe that the increasing frequency and severity of storms containing powerful, damaging winds is also the result of warming associated with greenhouse gas pollution.

So far, we have described the more obvious effects of global warming, but there are other equally important, if less visible, changes going on as well. Because of global patterns of air and water circulation and the way that sunlight hits the earth, areas closer to earth's poles are warming up more quickly than temperate latitudes. In the past, the Arctic contained vast areas where soils had long been frozen below the surface layer even in the summer. In some cases, these layers of *permafrost* were tens or even hundreds of feet deep. As earth's climate warms, however, the upper layers of permafrost are rapidly melting, and as this happens, they have begun to release not only significant amounts of carbon dioxide into the air, but also of methane, a greenhouse gas which, gram for gram, is twenty times as potent as carbon dioxide.[65] Within limits, this is a self-reinforcing cycle: higher temperatures lead to melting permafrost, which leads to the release of greenhouse gases and higher temperatures, which leads to more melting permafrost. Nor is the tundra permafrost the only large potential source of greenhouse gas methane. Many areas of the sea bottom in the Arctic contain large deposits of a material known as methane clathrate—essentially methane locked in ice crystals—in the sediment. Recent research shows that very large amounts of this methane is now being released into Arctic waters, and ultimately into the air, adding to the large amount already coming from melting permafrost.[66]

Our focus up to this point has been on the impact of global warming on the atmosphere, the tundra, and also on the droughts, fires and storms that have become more common. Yet even in areas where extreme events have not yet become more frequent, changes in average temperature are modifying habitats in ways that often make them less suitable for species that have long inhabited them. When this occurs quickly, as is happening today, some organisms' only workable survival strategy is to relocate to places where conditions resemble those they experienced in the past.[67] Yet even this response is available only to a limited range of living things. Some highly mobile creatures like migratory birds, fast-swimming ocean fish or large, wide-ranging predators like wolves or mountain lions may have no great difficulty in moving towards more suitable environments, although these will often be occupied by others of their kind with whom they will have to compete. But the vast majority of living things—plants, invertebrates, reptiles, amphibians, small mammals and others—are either rooted in place like trees or else have quite limited mobility. For them, relocation is either more difficult, or in practical terms, impossible.

In a warming world with higher temperatures and increasing droughts, habitat changes are likely to drive numerous creatures towards areas where climates are cooler or wetter than in their original range. For many, this may involve three kinds of movement: away from the Equator and towards the poles; closer to coastlines where there is generally more rainfall; or upwards in elevation where conditions are generally cooler and damper. Over the grand sweep of earth time, creatures have encountered situations like this before whenever geological changes or impacts from space produced relatively rapid changes in local climates. Today, however, the presence of human settlements and the widespread conversion of plant communities to agricultural land has made these movements much more difficult. Before agriculture, natural plant communities spread more or less contin-uously across the land, and if temperature change was not too rapid, plants could easily relocate northward or southward, perhaps at just some tens of feet a year. This was how forest trees in places like North America survived the relatively slow advance and retreat of Pleistocene glaciers. Today, however, vast areas of towns, roads and heavily modified farm and range land block the movement of natural vegetation in this way. The result is that many creatures will not be able to shift their ranges and so will simply die out locally once conditions in their current habitats become hotter or drier than they can tolerate. In mountainous regions, relatively short movements uphill can lead to much cooler habitats, although to ones that are likely to be full of well-established competitors. For creatures living on the lower slopes, this has long been a way of surviving warming temperatures. Hill-top and mountain-top species, however, have no place higher (or cooler) to retreat to, and some of these will also simply die out in coming decades as their previously cool to warm habitats become warm to hot ones.

Ocean Impacts

The challenges facing land creatures from technologically induced climate change are not too hard to imagine, but global warming also poses serious threats to life in the sea. Corals, for example, are important animals in clean, warm, tropical waters. Their ability to house themselves in growing limestone shells made out of minerals absorbed from seawater allows them, and the sponges and other creatures associated with them, to create some of the richest of earth's living habitats. Coral reefs cover only a relatively small area in warmer seas, but they are home to many tens of thousands of animals, plants and microbial species which together represent roughly one quarter of all living things in the oceans.[68] Before the industrial revolution, reefs saw relatively little disturbance, with the main human activity being the collection of fish and shellfish for food by subsistence foragers. By the twentieth century, however, all this began to change as tourism, recreation, mining, fishing with explosives and collections for the pet trade all began to take their toll, along with increasing damage due to water pollution from nearby coastal towns, industries and agriculture. Under this onslaught, reefs in many parts of the world began to deteriorate, with the greatest decline occurring in areas lacking environmental regulation, and where fishing with explosives, over-collecting of wildlife, dredging of channels, and damage from divers' boat anchors were most common. Then, in the latter part of the twentieth century, global warming itself began to directly impact these rich, unique, ancient marine communities.

Like so many other creatures in the natural world, reef-building corals live successfully by consistently cooperating with other living things. Most notably, they harbor large numbers of single-celled algae inside their limestone skeletons. The algae absorb waste carbon dioxide and nitrogen-rich wastes produced by the coral animals, while also carrying out photosynthesis. In this way, the algae feed themselves, but also supply some of their energy-rich carbohydrates to the corals that house them. Both organisms benefit from this relationship: the algae live relatively free from predation and have a built-in source of carbon dioxide and nutrients; the corals gain food and other important substances, some of which help them produce not only their limestone homes but also the remarkable pigments that make the reef one of earth's brightest and most colorful places.[69]

During the late twentieth century, as earth's atmosphere and land surface rapidly grew warmer, average ocean surface temperatures rose more than a degree Fahrenheit.[70] This may not sound like much, but in some tropical areas where other forces such as the El Niño/La Niña cycle of warming and cooling of Pacific surface waters were at work, the combined warming was enough to cause serious temperature stress for corals. When this happens, the algae they harbor often lose their ability to make chlorophyll; and if that occurs, the corals may no longer be able to maintain their cooperative relationship and instead may expel the algae. After doing so, some of the stressed coral animals generally die, while others lose their normal bright coloration and have difficulty forming new limestone shells, a condition in which they are described as *bleached*. Around the world in warm seas, coral bleaching events became more and more common from about 1980 on, and since then several widespread episodes have occurred damaging reefs all over the world. In locations where temperatures did not rise too high, or stay up for too long, many corals were able to recover. But where warming persisted for a significant period, bleaching was severe, with widespread coral death sometimes approaching 90 percent.[71]

Bleaching seems to be the most immediate threat faced by corals as a result of carbon dioxide pollution, but it is not the only one. As was mentioned earlier, as much as half of all the extra CO_2 produced by human activities has ended up dissolving in the sea. When this happens, carbonic acid is formed. It is not a strong acid, but its relentless build-up has been slowly changing the chemistry of the oceans in ways that will make life much more difficult for corals, mollusks and other marine creatures which form their shells from calcium carbonate. This substance—classroom chalk is one form of it—comes in two different natural mineral forms: calcite and aragonite. Both contain the same atoms, but the atoms are arranged differently in their crystals, giving them distinct physical and chemical properties. In the sea, an enormous range of creatures precipitates calcium carbonate out of seawater to build their shells and houses. Some, like sponges and echinoderms (starfish and their relatives), primarily use calcite; most mollusks (clams, snails, etc.) use both; and certain groups, especially corals, use only aragonite.[72]

In marine environments there is a complex interaction between calcium carbonate, the acidity of seawater and the formation of marine shells including corals. In the past, for at least hundreds of thousands of years, slightly alkaline seas have made it quite easy for living creatures to precipitate solid aragonite out of solution to build their bodies. But as carbon dioxide pollution has added more and more acid to seawater in recent centuries, the increased acidity has made it harder and harder for creatures to extract the solid aragonite that they need to build their skeletons. As this trend worsens in coming decades, a wide range of marine creatures will find it harder to live and grow, with corals being among the most susceptible to damage. And if this happens, not only will corals themselves decline, but the many thousands of living niches they create for fish, crabs and a wide range of other organisms will diminish along with them.[73]

At the present time, the best scientific estimate of how much hotter the earth is likely to get as we push atmospheric CO_2 up from its present level near 400 units to 500 units over coming decades is an additional 2.7°F (1.5°C). If this happens, earth may become warmer near the end of the twenty-first century than at any time during the past 400,000 years.[74] To put this in perspective, average temperatures rose just one degree Fahrenheit during the twentieth century. On the current trend, however, additional twenty-first century warming will be more than twice as great unless strong measures are taken to reduce the additional carbon dioxide and other greenhouse gases we continue to spew into the environment. That degree of warming over just a handful of decades will lead to ever more frequent episodes of coral bleaching and die-off, with the real possibility that by the end of this century most existing coral reef ecosystems on earth will have seriously declined or disappeared. If there were somehow a way to slow down this level of warming so that it occurred over, say, 200,000 years instead of 200, there is a good chance that corals and other organisms might successfully adapt. But 200 years is simply not enough time for life's natural adaptive mechanisms to be effective in the face of this degree of change. If coral reefs are allowed to disappear, it will not be just future human generations who will be impoverished; the overall potential for life on earth will also be diminished.[75]

Notes

1 B. B. K. Huat et al., 2011, A state of the art review of peat, *International Journal of the Physical Sciences* 6(8): 1988–1996, www.academia.edu/640528/State_of_an_Art_Review_of_Peat_General_Perspective.
2 Geology.com, Coal, http://geology.com/rocks/coal.shtml.
3 Wikipedia, Neolithic flint mines of Spiennes, https://en.wikipedia.org/wiki/Neolithic_flint_mines_of_Spiennes.
4 Wikipedia, Steam shovel, https://en.wikipedia.org/wiki/Steam_shovel; Wikipedia, Panama Canal, https://en.wikipedia.org/wiki/Panama_Canal.
5 Wikipedia, Steam shovel, https://en.wikipedia.org/wiki/Steam_shovel.
6 Kansas Sampler Foundation, Big Brutus, www.kansassampler.org/8wonders/8wonder-sofkansas-view.php?id=20.
7 Wikipedia, Bingham Canyon mine, https://en.wikipedia.org/wiki/Bingham_Canyon_Mine.
8 Appalachian Voices, Mountaintop removal 101, http://appvoices.org/end-mountaintop-removal/mtr101; Wikipedia, Mountaintop removal mining, https://en.wikipedia.org/wiki/Mountaintop_removal_mining.
9 MiningFacts.org, What is acid rock drainage?, www.miningfacts.org/Environment/What-is-acid-rock-drainage.
10 Ibid.
11 Wikipedia, Rio Tinto (river), https://en.wikipedia.org/wiki/Rio_Tinto_%28river%29; Science Buddies, Acids, bases, & the pH scale, www.sciencebuddies.org/science-fair-projects/project_ideas/Chem_AcidsBasespHScale.shtml.
12 G. Watson, 2012, Polluted legacy: repairing Britain's damaged landscapes, *BBC News* (June), www.bbc.com/news/uk-england-derbyshire-17315323.
13 Wikipedia, Coal seam fire, https://en.wikipedia.org/wiki/Coal_seam_fire.
14 SourceWatch, Coal fires, www.sourcewatch.org/index.php/Coal_fires.
15 International Energy Agency, Coal, www.iea.org/topics/coal.
16 Wikipedia, Environmental impact of the coal industry, https://en.wikipedia.org/wiki/Environmental_impact_of_the_coal_industry.
17 T. Gabriel, 2014, Ash spill shows how watchdog was defanged, *New York Times* (February 28), www.nytimes.com/2014/03/01/us/coal-ash-spill-reveals-transformation-of-north-carolina-agency.html?_r=0.
18 US Geological Survey, 1997, Radioactive elements in coal and fly ash: abundance, forms, and environmental significance, http://pubs.usgs.gov/fs/1997/fs163-97/FS-163-97.html; National Institutes of Health, Tox town: dioxins, https://toxtown.nlm.nih.gov/text_version/chemicals.php?id=12.
19 R. D. Brook et al., 2004, Air pollution and cardiovascular disease, *Circulation* 109(21): 2655–2671.
20 J. Kirkup, 2004, Percivall Pott, *Oxford Dictionary of National Biography*, www.oxforddnb.com.
21 C. Hamlin, 2008, Robert Angus Smith, *Oxford Dictionary of National Biography*, www.oxforddnb.com.
22 Ibid.
23 Power Plants, Flue-gas desulfurization, http://powerplantstechnology.blogspot.com/2010/08/flue-gas-desulfurization.html.
24 Staffordshire University, Medieval Pollution, www.staffs.ac.uk/schools/sciences/environment/GreatFog/fog2.html.
25 Wikipedia, Fumifugium, https://en.wikipedia.org/wiki/Fumifugium.
26 Royal Geographical Society, London's killer smog, www.rgs.org/OurWork/Schools/Teaching+resources/Weather+and+climate+resources/Key+Stage+Three/Londons+killer+smog.htm.
27 UK Met Office, The great smog of 1952, www.metoffice.gov.uk/education/teens/case-studies/great-smog.

28   Legislation.gov.uk, Clean Air Act of 1956, www.legislation.gov.uk/ukpga/Eliz2/4-
      5/52/enacted; Wikipedia, Clean Air Act of 1956, https://en.wikipedia.org/wiki/
      Clean_Air_Act_1956.

29   E. Peterman, 2009, A cloud with a silver lining: the killer smog in Donora, 1948,
      http://pabook2.libraries.psu.edu/palitmap/DonoraSmog.html; Wikipedia, 1948
      Donora smog, https://en.wikipedia.org/wiki/1948_Donora_smog.

30   United States Senate Committee on Environment and Public Works, Clean Air Act,
      www.epw.senate.gov/envlaws/cleanair.pdf; USEPA, Overview of the Clean Air Act
      and air pollution, www.epa.gov/clean-air-act-overview.

31   R. Wolf, 2014, Supreme Court limits greenhouse gas regulations, *USA Today* (June 23),
      www.usatoday.com/story/news/nation/2014/06/23/supreme-court-greenhouse-
      gas/8567453.

32   USEPA, EPA's Budget and spending, www2.epa.gov/planandbudget/budget.

33   US Government Accountability Office, 2011, Information on tall smokestacks and their
      contribution to interstate transport of air pollution, www.gao.gov/products/GAO-11-
      473; Wikipedia, List of tallest chimneys in the world, https://en.wikipedia.org/
      wiki/List_of_tallest_chimneys_in_the_world.

34   USEPA, Environmental effects of acid rain, www.epa.gov/region1/eco/acidrain/
      enveffects.html; Wikipedia, Acid rain, https://en.wikipedia.org/wiki/Acid_rain.

35   USDA Forest Service, Red spruce reviving in New England, but why?,
      www.nrs.fs.fed.us/news/release/reviving-red-spruce.

36   B. J. Skinner and S. C. Porter, 1987, *Physical Geology*, Wiley, pp. 597–605.

37   Wikipedia, Kuwaiti oil fires, https://en.wikipedia.org/wiki/Kuwaiti_oil_fires.

38   Texas State Historical Association, Spindletop oilfield, www.tshaonline.org/handbook/
      online/articles/dos03.

39   Ibid.

40   Wikipedia, List of oil refineries, https://en.wikipedia.org/wiki/List_of_oil_refineries.

41   National Public Radio, Baton Rouge's corroded, over-polluting neighbor: Exxon-
      Mobil, www.npr.org/2013/05/30/187044721/baton-rouge-s-corroded-overpolluting-
      neighbor-exxon.

42   F. A. Wichmann et al., 2009, Increased asthma and respiratory symptoms in children
      exposed to petrochemical pollution, *Journal of Allergy and Clinical Immunology* 123(3):
      632–638.

43   Wikipedia, Oil tanker, https://en.wikipedia.org/wiki/Oil_tanker.

44   Shipwreck Log, Atlantic Empress, www.shipwrecklog.com/log/history/atlantic-
      empress.

45   Exxon Valdez Oil Spill Trustee Council, Questions and answers about the spill,
      www.evostc.state.ak.us/index.cfm?FA=facts.QA; Wikipedia, Exxon Valdez oil spill,
      https://en.wikipedia.org/wiki/Exxon_Valdez_oil_spill.

46   US NOAA, Office of Response and Restoration, 2014, A final farewell to oil tankers
      with single hulls, http://response.restoration.noaa.gov/about/media/final-farewell-oil-
      tankers-single-hulls.html.

47   US NOAA, Office of Response and Restoration, Deepwater Horizon oil spill,
      http://response.restoration.noaa.gov/deepwater-horizon-oil-spill?page=2; US NOAA,
      Gulf spill restoration, www.gulfspillrestoration.noaa.gov/2016/04/trustees-settle-with-
      bp-for-natural-resource-injuries-to-the-gulf-of-mexico; Wikipedia, Deepwater
      Horizon oil spill, https://en.wikipedia.org/wiki/Deepwater_Horizon_oil_spill.

48   T. Pering, 2009, The history and philosophy of glaciology, www.volcano-
      blog.com/uploads/6/9/7/6/6976040/the_history_and_philosophy_of_glaciology.pdf;
      Wikipedia, Ice age, https://en.wikipedia.org/wiki/Ice_age.

49   S. Arrhenius, 1896, On the influence of carbonic acid in the air upon the temperature
      of the ground, *Philosophical Magazine and Journal of Science* 41: 237–276,
      www.rsc.org/images/Arrhenius1896_tcm18-173546.pdf; Wikipedia, Svante Arrhenius,
      https://en.wikipedia.org/wiki/Svante_Arrhenius.

50 K. A. Zimmermann, 2013, Temperature: facts, history & definition, www.livescience.com/39841-temperature.html.

51 UK Met Office, 2014, Overview of the Met Office, www.metoffice.gov.uk/news/in-depth/overview#origins.

52 US NOAA, History of the national weather service, www.weather.gov/timeline.

53 Wikipedia, Temperature record, https://en.wikipedia.org/wiki/Temperature_record.

54 University of Copenhagen, Centre for Ice and Climate, Niels Bohr Institute, 2014, The history of Danish ice core science, www.iceandclimate.nbi.ku.dk/about_centre/history.

55 Wikipedia, Oxygen isotope ratio cycle, https://en.wikipedia.org/wiki/Oxygen_isotope_ratio_cycle.

56 Scripps Institute of Oceanography, Scripps CO_2, CO_2 concentration at Mauna Loa observatory, Hawaii, http://scrippsco2.ucsd.edu.

57 D. M. Etheridge et al., 1996, Natural and anthropogenic changes in atmospheric CO_2 over the last 1000 years from air in Antarctic ice and firn, *Journal of Geophysical Research* 101(D2): 4115–4128, www.acoustics.washington.edu/fis437/resources/Week%2010/Etheridge%20et%20al.%201996.pdf.

58 Wikipedia, Carbon dioxide in Earth's atmosphere, https://en.wikipedia.org/wiki/Carbon_dioxide_in_Earth's_atmosphere.

59 USEPA, Climate change: basic information, www.epa.gov/climatechange/basics.

60 US Geological Survey, Retreat of glaciers in Glacier National Park, www.usgs.gov/centers/norock.

61 National Snow and Ice Data Center, Quick facts on ice sheets, https://nsidc.org/cryosphere/quickfacts/icesheets.html.

62 Ibid.

63 UK Met Office, The heatwave of 2003, www.metoffice.gov.uk/learning/learn-about-the-weather/weather-phenomena/case-studies/heatwave; Wikipedia, 2003 European heat wave, https://en.wikipedia.org/wiki/2003_European_heat_wave.

64 Wikipedia, 2010 Russian wildfires, https://en.wikipedia.org/wiki/2010_Russian_wildfires.

65 USEPA, Overview of greenhouse gases: methane emissions, http://epa.gov/climatechange/ghgemissions/gases/ch4.html.

66 Ibid.; N. Shakhova et al., 2014, Ebullition and storm-induced methane release from the East Siberian Arctic Shelf, *Nature Geoscience* 7: 64–70.

67 A. Murray, 2016, First-ever online B.C. Breeding Bird Atlas finds many bird species are on the move, *The Georgia Straight* (May 20), www.straight.com/news/702641/anne-murray-first-ever-online-bc-breeding-bird-atlas-finds-many-bird-species-are-move.

68 Smithsonian Ocean Portal, Corals and coral reefs, http://ocean.si.edu/corals-and-coral-reefs; Wikipedia, Coral reef, https://en.wikipedia.org/wiki/Coral_reef.

69 US NOAA, National Ocean Service, 2014, Corals, http://oceanservice.noaa.gov/education/tutorial_corals/welcome.html.

70 J. Hansen et al., 2006, Global temperature change, *PNAS* 103(39): 14,288–14,293, www.pnas.org/content/103/39/14288.full.

71 L. Burke et al., 2011, Reefs at risk revisited, www.wri.org/sites/default/files/reefs_at_risk_key_findings.pdf.

72 R. A. Feely et al., 2004, Impact of anthropogenic CO_2 on the $CaCO_3$ system in the oceans, *Science* 305(5682): 362–366; Wikipedia, Ocean acidification, https://en.wikipedia.org/wiki/Ocean_acidification.

73 Scientific Committee on Oceanic Research, 2009, Report of the Ocean Acidification and Oxygen Working Group, www.scor-int.org/OBO2009/A&O_Report.pdf.

74 T. F. Stocker et al. (eds.), IPCC climate change 2013: the physical science basis, www.ipcc.ch/report/ar5/wg1; O. Hoegh-Guldberg et al., 2007, Coral reefs under rapid climate change and ocean acidification, *Science* 318(5857): 1737–1742.

75 USEPA, Future climate change, www3.epa.gov/climatechange/science/future.html; A. Schmittner et al., 2011, Climate sensitivity estimated from reconstructions of the Last Glacial Maximum, *Science* 334(6061): 1385–1388.

8

NUCLEAR POWER AND RENEWABLE ENERGY

Nuclear Power

The invention of Newcomen's steam engine around 1710 marked the first time that humanity's old friend *fire* could readily be harnessed to generate artificial power. Watt and Boulton engines that had replaced Newcomen's design by the last decades of the eighteenth century combined the force of steam at slightly above atmospheric pressure with the vacuum generated by a condenser to move a piston within a cylinder. This, in turn, allowed some of the energy produced by the fire in the boiler to be transformed into mechanical force. A century or so later, when the first successful rotary steam engine was developed, the working principles remained essentially the same, even though the energetic steam was now at much higher pressure and was used to spin a turbine rather than to move a piston lengthwise within a cylinder. People had known about steam for thousands of years, but until Newcomen, Watt and Parsons, they had never been able to effectively harness it. Towards the end of the nineteenth century, in contrast, physicists were amazed to discover a completely new source of energy, one that appeared to emanate spontaneously from inside the atoms of some chemical elements.

On earth, the vast majority of the atoms that make up matter are physically stable—they are not radioactive. They constantly interact with one another to maintain, form or break chemical bonds consistent with the rules of chemistry and the influence of surrounding physical conditions like heat and pressure. Relatively simple chemical interactions leading to new combinations of atoms (molecules) occur all the time, although in the absence of life, large and very large molecules are extremely rare. And in the non-living world there is also nothing like the tremendously complicated chains of linked chemical reactions needed to keep even the simplest bacterial cells alive.

Naturally occurring radioactivity and the heat it released was first discovered in the 1890s, and over the following decades, its origins in the nuclei of unstable atoms gradually became clear. As this understanding emerged, thoughtful scientists began to speculate about whether nuclear energy could potentially form the basis of a new kind of power technology. If heat from a wood or coal fire could be used to boil water to run a turbine, why couldn't the heat produced by radioactive decay do the same? And if such *atomic heat* could be produced in large amounts in a controlled way, perhaps combustion could be dispensed with completely in favor of a slowly decaying mass of radioactive material that created enough heat to continuously boil water. By the second half of the twentieth century, this concept was developed as the basis of the civilian nuclear power industry.

The fundamental scientific discoveries underlying the development of atomic energy were made between 1899 and 1945 by academic physicists and chemists like Ernest Rutherford, John Cockcroft, Ernest Walton, Niels Bohr, Enrico Fermi, Lise Meitner, Otto Hahn and Fritz Strassmann. Their basic research had led to a model of the atom in which most of its volume was filled with a lightweight cloud of electrons, while most of its mass was concentrated in a tiny volume in a central nucleus. Based on this understanding, for example, Rutherford had demonstrated in 1917 that nitrogen could be transformed by bombardment with helium nuclei (alpha particles) into oxygen, the next heavier element in the Periodic Table, thus fulfilling for the first time the alchemist's dream of actually transforming one element into another. As late as the 1930s, however, physicists continued to believe that while it was possible to *add* sub-atomic particles to atoms to create heavier elements—a process accompanied by some release of radiation—it was impossible to *split* the larger nuclei of heavy elements into fragments, a process which eventually came to be known as *nuclear fission*.

In 1938, however, Hahn and Strassmann bombarded tiny quantities of uranium (U) with neutrons and found that some of the by-products produced were roughly half the mass of the uranium atoms they had started with. Then Meitner and Otto Frisch, who as refugees from Nazi Germany had just escaped to Sweden, provided the first accurate explanation of Hahn and Strassmann's experiments when they suggested that some of the uranium atoms had in fact been split apart following their absorption of neutrons. Then they went on to conclude, both from their analysis of Hahn and Strassmann's data, and from their own extensive calculations, that the resulting fission products weighed significantly less than the uranium starting materials. From this they reasoned that the lost mass represented an enormous potential source of energy, one that they could estimate using Albert Einstein's famous mass–energy equivalency equation, $E = mc^2$, which stated that the loss of mass yields an amount of energy in proportion to the speed of light squared. Hahn and Strassmann had experimented with miniscule samples of radioactive materials. But after the analyses of Meitner, Frisch and other researchers, it soon became clear to physicists that complete nuclear fission of a given weight of suitable uranium could liberate roughly *three million times as much energy* as complete burning of an equal mass of coal.[1]

As research continued during the late 1930s, it was also discovered that only a relatively uncommon isotope of uranium—U-235—could easily undergo this fission reaction. Atoms of that isotope naturally emitted neutrons, and when as little as 50 kilograms (110 pounds) of U-235—a quantity the size of a large brick—was brought together as a pure substance, it could spontaneously initiate an on-going nuclear fission reaction. Once this occurred, the splitting of each subsequent uranium atom into atoms of smaller elements released not only large amounts of heat, but also a variety of other kinds of radiation including additional neutrons, which, in turn, would split other uranium atoms, keeping the fission process going. Such a *chain reaction* could then continue indefinitely, ending only when most of the reactive uranium-235 had broken down to other elements.

The Atomic Bomb

World War II began in 1939, and on both sides of that global conflict top secret research soon started on ways of using the newly discovered nuclear reactions to produce more powerful explosives. One early discovery of this classified research was a previously unknown, super-heavy metal called plutonium (Pu). Besides being radioactive, it had other unusual chemical and physical properties, among which was the ability of relatively small quantities of it to readily undergo nuclear fission. Once this was understood, the same kinds of brilliant theoretical insights and mathematical calculations that had led to Hahn and Strassmann's groundbreaking experiments soon made it just as clear that it was possible to build a device in which modest quantities of U-235—or even smaller amounts of Pu-239—could be brought together in such a way as to trigger an even faster chain reaction. Under these conditions, runaway fission would occur, generating so much heat and physical expansion that it would behave like an enormous explosive—a *nuclear bomb*.

The result of all this was that the first practical technology utilizing significant quantities of radioactive material was the enormously destructive atomic (fission) bomb. Following a secret, wartime crash program, the first such weapon utilizing plutonium was assembled and detonated in the United States on July 16, 1945 in an atmospheric explosion in the state of New Mexico. And even as that test was being carried out, a summit meeting of the World War II Allied Powers in Potsdam, Germany was busy drawing up terms of surrender for Imperial Japan—Germany had already surrendered. These were published on July 26 as the Potsdam Declaration, although the Japanese government quickly rejected them. Then, on August 6, just twenty days after the very first ever fission bomb test and fearful of having to stage a land invasion of the Japanese home islands costing countless lives on both sides, the United States detonated the world's first uranium nuclear bomb over Hiroshima. It completely destroyed the city, killed roughly 100,000 people, and released significant quantities of radioactive debris into the global environment. Immediately afterwards, the US Government made a public announcement that a new type of super-bomb had been employed in the

Hiroshima attack, and that further attacks would follow unless the terms laid out at Potsdam were accepted. There was, however, no quick reaction from the Japanese government, and on August 9, a second nuclear bomb, this one based on runaway fission of plutonium, was exploded over Nagasaki, leveling that city, killing tens of thousands of additional people, and further polluting the atmosphere with radioactive materials.[2]

By August 1945, tens of millions of people had already died in World War II, one of the most destructive episodes in human history. During that conflict's countless battles, small mountains of harmful lead and other toxic substances had been dispersed from explosives, bullets, fuels, batteries, industrial solvents, etc., while the total number of victims and the damage to social infrastructure—houses, factories, roads, bridges, schools and the like—was nearly incalculable. In addition, there was also enormous, and largely unrecorded, damage to wild creatures and natural systems, especially in active theatres of war in Europe, North Africa and the Asia-Pacific region. Yet, despite all that horror, bloodshed and destruction, it was only in the very final days of the war, with the explosion of the first nuclear weapons, that human technology introduced an entirely new category of danger to the living world, the threat of poisoning, mutations and cancer from exposure to artificially produced radioactivity.

★ ★ ★

If the bloody end of World War II had led to lasting peace and a halt to further development of nuclear weapons, the dropping of the fission bombs on Japan would probably be viewed as just another highly tragic episode in human history, rather than as a watershed change in the impact of human technology on the natural world. Unfortunately, this did not turn out to be the case. The Japanese surrender in August 1945 was followed by only the briefest of lulls in the advance of nuclear technologies, which thereafter developed in several directions, one of which involved continued production of bigger fission, and later hydrogen (fission/fusion) bombs. The result was that between 1945 and 1963, nearly 550 atomic and hydrogen bombs were tested in the open air, mostly by the United States and the Soviet Union, freely dispersing their radioactive and non-radioactive debris into earth's environment.[3]

Atmospheric bomb testing released a wide range of radioactive isotopes. For plutonium, to take just one example, an estimated eleven tons (10,000 kilograms) was put into earth's environment, the vast majority of which was Pu-239, which has a half-life of 24,065 years.[4] Given its extremely slow rate of decay, essentially all of that plutonium is still present and still radioactive. Because of the truly global dispersal of radioactive debris like plutonium from atmospheric testing, there was no wave of localized deaths in either people or wildlife. But global background radiation certainly rose from those releases, at least slightly increasing the likelihood of mutations and cancer all around the world, in people and in other living things. Only in 1963, after nearly two decades of atmospheric testing, did a number of nations finally sign an international treaty banning open air explosions. This ended the outdoors test programs of the United States, the Soviet Union and Britain, although France continued to explode nuclear bombs in the atmosphere until 1974, and China until 1980.[5]

A second, relatively minor, application of nuclear technology beginning in the 1950s was in ship's propulsion, beginning with the first nuclear submarine, the *Nautilus*. It was followed by the first atomic icebreaker, the *Lenin*, and by a nuclear-powered freighter, the *Savannah*, launched in 1959 to demonstrate peaceful uses of nuclear power. In 1961, the first nuclear aircraft carrier, the *Enterprise*, was launched. Given the overall costs and hazards, however, it was eventually realized that nuclear propulsion was not a viable technology for use in ordinary civilian ships, and most of the roughly 150 atomic powered vessels that have been built over the years have been warships, particularly nuclear submarines, whose combustion-free power plants allow them to remain submerged for months on end.[6]

Beyond military applications, the most important path taken by atomic technologies was the creation of nuclear power stations for the production of electricity. In these plants, the heat of *controlled* nuclear fission boils water to run steam turbines which drive electrical generators. This basic technology was first demonstrated by a plutonium breeder reactor in Idaho in 1951, while a small, dedicated nuclear electricity generating station came into service at Obninsk in the Soviet Union in 1954.[7] By 1956, a commercial-scale nuclear electric plant had also started operating at Sellafield in Great Britain, although it was simultaneously producing weapons grade plutonium for nuclear bombs.[8] Throughout the rest of the twentieth century, hundreds of nuclear power stations were built, and by the first decade of the twenty-first century, about 440 plants around the world were generating roughly one seventh of the world's electric power, along with the majority of its newly produced nuclear waste.[9]

Nuclear Waste and Debris

During the twentieth century, physics experiments and nuclear technologies produced hundreds of different radioactive isotopes either by concentrating them from minerals (e.g., U-238 and U-235) or by creating them in nuclear explosions, nuclear power reactors, or particle accelerators. Each different radioisotope proved to have its own pattern of radioactive emissions, its own decay rate, and its own set of interactions with living tissues. These properties, along with the isotope's chemical reactivity, and with the quantity and pattern of its creation and release, determined the degree and duration of the damage it could do. Radiation itself was found to take several different forms: alpha radiation, which was made up of helium nuclei; beta radiation, which consisted of electrons; gamma and X-ray electromagnetic radiation, which are not altogether different from light, but are of much higher energy; and neutrons, the sub-atomic particles which normally contribute significantly to the mass of atoms and which form the basis for the existence of different isotopes. The particles and waves present in radiation can exist at different energy levels, and usually the more energetic they are, the more damage they can do to living cells. Many isotopes were produced only in low concentrations, and some decayed so quickly to stable end-products that short-term storage was all that was needed to render them harmless. A few, though, proved to be radioactive enough, common enough, or long-lasting enough to present serious dangers to people and other living things.

The processed *enriched uranium* used in most power plants and in many nuclear bombs is a mixture of U-238 and U-235 which contains significantly more U-235 than is normally present in uranium ores. Enrichment is carried out because atoms of U-238 do not undergo nuclear fission. U-235 atoms, on the other hand, are capable of undergoing fission in both bombs and reactors. During uranium fission, aside from the release of very large amounts of energy, a flood of sub-atomic particles is also produced which transform the many different atomic fragments into as many as 370 different isotopes belonging to many different chemical elements.[10] As a result, both radioactive wastes from reactors—most notably spent nuclear fuel—and dust and debris from atomic explosions contain a very wide range of radioactive materials whose complexity required several decades of research in the twentieth century to unravel. Let us consider one or two important examples.

Radioiodine

Among the common, short-lived radioactive by-products of uranium fission, radioactive iodine is probably the most dangerous. Long ago, in the course of evolution, as animals with backbones were evolving in the oceans, atoms of the element iodine came to be utilized along with the amino acid tyrosine to synthesize the vertebrate thyroid hormones—key signaling molecules designed to regulate metabolism. In mammals, these are produced and stored as needed in the thyroid gland located in the neck. In other vertebrates, thyroid tissue is present, but it is not located in a single gland. For this reason, nearly all the iodine that mammals encounter in their diet, either stable or radioactive, ends up being concentrated in their thyroid tissues. Because it is essential for child development and growth, iodine is also a normal constituent of mammalian milk, and radioiodine can get into the human body by consuming dairy products produced by cows that have eaten contaminated grass or hay.[11]

In the natural environment, most iodine exists as non-radioactive I-127, but radioactive iodine-131, which has the same exact chemical behavior as I-127, is routinely produced by uranium fission with roughly a 3 percent yield. It has a short half-life of eight days, and thus will have almost all decayed after about three months. But if it is released into the environment shortly after being created, it will be readily taken up by animals and people, delivering significant doses of beta and gamma radiation and dramatically increasing the incidence of thyroid cancer, especially in children. This first came to be understood in the aftermath of bomb testing in the twentieth century, after several hundred atmospheric nuclear tests between 1945 and 1980 released significant amounts of it.

Nuclear test explosions in the atmosphere spread their invisible contamination over a large area, resulting in somewhat elevated cancer risks for very large populations, although in most cases they did not immediately cause significant numbers of deaths. If public health preparations are made and adequate warnings of a release are given, people can swallow harmless iodine salts in large enough quantities to block the uptake of any radioactive iodine. But in the mid-twentieth-century era of atmospheric nuclear tests, the dangers were not fully understood, and neither warnings nor medication were made available outside the immediate test areas. Although the radioactive fallout from testing was widely dispersed, hundreds of millions of people and billions of organisms were still exposed. In 1997, the US National Cancer Institute estimated that in the United States alone, where 86 atmospheric and 14 partially contained underground nuclear tests were conducted in Nevada between 1951 and 1962, enough radioactive iodine was released to cause 49,000 additional cases of thyroid cancer in the American population over the lifetimes of all those exposed.[12] No good data are available on the impact of this fallout on wild populations of vertebrates who use similar thyroid hormones.[13]

Cesium 137

Cesium-137 is another dangerous, moderately long-lived by-product of uranium fission, which poses a significant risk of cancer. It emits beta and gamma radiation, has a six percent yield in fission reactors and bombs, and possesses a half-life of slightly more than thirty years. It is very mobile in the environment, readily enters the food chain, and easily ends up in things that people eat. Its gamma radiation is fairly penetrating, and so unlike radioiodine, external exposure to it can also be harmful. Once taken up into the body, it is distributed like potassium, with higher concentrations in muscle and lower amounts in bone.[14]

Uranium and Plutonium

The overall abundance of uranium in the earth's crust is roughly two parts per million. This makes it uncommon, but not extremely rare—gold, for example, is hundreds of times less common. Of all that uranium, however, less than one percent is the isotope U-235, which readily undergoes fission. During the 1940s, the desperate psychology of wartime turned basic physics research into an applied weapons development program emphasizing the design of bombs and the production of fissionable material to use in them. Uranium-235 was the first material focused on, and in the United States, an enormous, secret, gaseous-diffusion plant was built at Oak Ridge, Tennessee in which thousands of workers toiled to separate U-235 from the more common U-238. Eventually, this effort produced enough material for the U-235 bomb that destroyed Hiroshima, Japan in August 1945. In the early 1940s, of course, nuclear weapons were an entirely new and unproven technology, and so in addition to the massive efforts to enrich uranium, the US also undertook a crash program to produce a second fissionable material—plutonium—which could also be used to make bombs.

Plutonium had first been identified in 1941, after being produced in a high-energy physics experiment in which U-238 had been bombarded with nuclei of deuterium atoms (deuterium is an unusual, heavy isotope of hydrogen). In the decades following its discovery, miniscule traces of it were also detected in nature, probably resulting from earlier geological nuclear reactions of uranium, although natural plutonium is so rare as to not even be included in many lists of the elements that make up the earth's crust. During World War II, however, one plutonium isotope, Pu-239, became the subject of intense, secret wartime research after it became clear that along with U-235, it was capable of undergoing runaway nuclear fission and could therefore be used in bombs. To produce it, the United States built a number of specialized nuclear reactors in several locations, including the vast military installation in Hanford, Washington, where the relatively abundant U-238 was bombarded with neutrons to produce Pu-239. The result was the raw material for both the very first nuclear weapon that was tested in New Mexico in July 1945, and for the bomb which destroyed the Japanese city of Nagasaki on August 9, 1945.[15]

World War II came to an end almost immediately after the Nagasaki bombing, although it was soon followed by the Cold War era of political and military tension between the Soviet Union and the western allies. In 1949, the Soviets exploded their own fission bomb and soon both sides were producing large quantities of plutonium in secret armaments plants for weapons purposes. This was the source of approximately half of the estimated 1,300 metric tonnes (2.86 million pounds) of plutonium now present on the earth.[16] Most of the other half of the world's plutonium has come from civilian reactors in which uranium fission is used to produce heat for power generation. Hundreds of such nuclear-electric plants were constructed all around the world in the second half of the twentieth century, and almost all of them produce some plutonium in the course of their normal operation.

When atoms of U-235 split, they release neutrons along with a range of other fission products. Some of these neutrons serve to break apart other U-235 atoms creating a *chain reaction*. Others, however, are instead captured by atoms of U-238, which do not fission, but which then undergo radioactive decay in two steps to Pu-239. About half of the Pu-239 produced in this way eventually also undergoes fission inside the fuel rods where it is created, a reaction which generates roughly a third of the useful heat produced in the reactor, and also produces some of the reactor's nuclear fission products.[17] The remaining Pu-239 ends up in the spent fuel rods when the rods are retired after three to six years in the reactor. Every year, in this way about *70 tonnes* of Pu-239 are produced by civilian nuclear power plants around the world. (Other isotopes of plutonium are also produced in a uranium power reactor, and all are radioactive and dangerous; but only Pu-239 is readily usable for bomb-making, and so is the by-product that is of greatest concern.) At times in the past, Pu-239 from power reactors was painstakingly separated out from spent uranium fuel rods, after which it could either be mixed into new fuel rods to make uranium–plutonium (mixed oxide) nuclear fuel, or else sent directly to bomb factories. But since it is easier to make new fuel rods from uranium ore, and to produce military Pu-239 directly in special reactors, nowadays the plutonium in spent uranium fuel rods is simply left in place in the spent fuel rods where it becomes a long-lived component of *high level nuclear waste.*

The chief radiation produced by Pu-239 consists of alpha particles. Outside the body, its radioactivity presents only a low-level health threat because most of its radiation does not penetrate very far. If, however, even relatively small amounts of it get into the blood or the lungs, our bodies (and those of other vertebrates) are very slow at eliminating it. When that happens Pu-239 remains quietly in place for many years localized primarily in lung, liver, lymph node, and bone tissues, where it increases the chance of developing cancer in those organs. Because it is most dangerous if it is inhaled or enters a wound, it also features prominently in concerns about its use by terrorists or rogue governments as part of a *dirty bomb*, in which radioactive material would be combined with conventional explosives in a weapon designed to be detonated in a crowded city. Fortunately, most of the several tons of plutonium that was released into earth's atmosphere by decades of atmospheric nuclear testing was in an insoluble form; it was also so widely dispersed that few if any health effects have ever been detected from it—much of it has ended up in ocean sediments.[18]

Radioactive Waste and the Future

The age of nuclear power is less than a century old, yet even in the mid-twentieth century, the possible use of nuclear weapons more than once brought the world to the brink of catastrophe. Overall, military nuclear technologies have left a significant legacy of pollution in the form of bomb fallout, stockpiles of plutonium, heavily polluted manufacturing sites, and contaminated zones around places where nuclear plant accidents occurred. Two of the most notable of the manufacturing sites are near Ozyorsk (Chelyabinsk) in Russia, and at Hanford, Washington in the United States. Though much talked about, world nuclear disarmament has not made much progress, and the list of states building nuclear weapons has slowly increased over time, with North Korea being the most recent addition in 2006. Broadly speaking, however, military nuclear technologies have not been rapidly expanding in recent decades, and the main area in which atomic energy remains a growth industry is in power plants for the generation of electricity. As of the early twenty-first century, for example, more than 400 nuclear power stations for generating electricity had been built around the world along with more than 100 smaller nuclear power plants on naval vessels.[19] Until a serious nuclear accident in Japan in 2011, many dozens of new nuclear-electric plants were being planned, and a few countries are still moving ahead with planned construction.[20]

Over the decades since the debut of nuclear technologies in the 1940s, a great deal has been learned about the complexities involved in producing, using and disposing of radioactive materials. Knowledge has also been gained about the dangers involved in extracting and refining nuclear fuels, of turning them into power plant fuel rods or weapons' components, of handling and transporting them, and of disposing of them at the end of their useful life. In general, civilian nuclear power plants are far more expensive to build than fossil fuel power plants, although once built, they are cheaper to run because in the short run at least, nuclear fuel is much less expensive per unit of heat generated than combustible fuels. In addition, nuclear electricity contributes relatively little to human global warming because it produces very little carbon dioxide. Once the end of their useful life is reached, however, nuclear power plants cost vastly more than fossil fuel plants to shut down largely due to the hazardous radioactivity that remains in their spent nuclear fuel and other radioactive components. The latter includes parts of the plant machinery and buildings. Many decommissioned nuclear plants also end up as medium-term storage facilities for the high-level radioactive wastes that were generated on site, and this requires expensive, open-ended guarding and monitoring to ensure that radioactive materials do not fall into the hands of criminals or terrorists.

Nuclear power's ability to reliably generate significant amounts of electricity while producing only low levels of greenhouse gas emissions clearly represents an important advantage. But the technology also brings with it two very important, and so far insurmountable challenges. One is the continuing dangers of the diversion of nuclear material to weapons of mass destruction. The other is the ever-accumulating stockpiles of highly dangerous radioactive wastes. Every day around the world, non-nuclear industries produce large amounts of a wide range of hazardous and even deadly non-radioactive chemicals (see Chapter 4). But almost without exception, reliable techniques, including high-temperature burning, are capable of transforming all of these products into no-risk/low-risk substances that safely can be disposed of. Radioactive wastes, on the other hand, fall into an entirely different category. The hazards they represent are rooted in the deeper, physical properties of atoms, rather than in the relatively superficial interactions of electrons that form the basis of chemical reactions. For this reason, in assessing the overall impact of nuclear technologies on the environment, it is essential to consider the *long-term* implications of the continually growing piles of nuclear wastes, materials which are usually categorized as high level, intermediate-level, or low-level, based primarily on the nature of their radioactivity and the hazards they present. High-level wastes—the most radioactive—are of greatest concern.

By the first decade of the twenty-first century, the International Atomic Energy Agency (IAEA) estimated that there was already 830,000 cubic meters (1,085,000 cubic yards) of high-level waste around the world.[21] To place that amount in perspective, it has been separately estimated that since the start of mining in prehistoric times, a grand total of just 8,500 cubic meters of gold has been produced by all societies on all continents. If we use this yardstick, current stockpiles of high-level nuclear waste amount to roughly one hundred times as much material as all the gold in the world.[22] To be sure, many non-radioactive, toxic industrial wastes exist in greater volumes than this, but none poses anything like the profound, long-lasting public health and safety challenges of high-level nuclear wastes.

Plutonium-239 offers a critical example. It is routinely produced in significant amounts in nuclear power plants (in addition to the large stocks left over from weapons production), is moderately to very dangerous to human and animal health, and can be stolen and diverted to make fission bombs and *dirty* conventional weapons. Most troubling of all, its half-life of 24,065 years means that it will have to be guarded against leakage and illicit diversion for weapons for as long as *a quarter of a million years*, during which time it will have to be kept completely away not only from people, but also from natural cycles involving groundwater and atmospheric gases and dust. This period of time is *fifty times longer than all of recorded human history*, and it remains highly questionable whether such long-term safeguarding can actually be achieved.

Over the last several decades, billions of dollars have been spent on projects in a number of countries designed to bury high-level radioactive wastes in geologically stable locations. As of this writing, however, no fully workable long-term disposal has ever been accomplished. Instead, ever greater quantities of spent fuel rods and other high-level wastes continue to accumulate at power plants and other nuclear installations around the world, with no clear plan in place for their permanent disposal. In the past, human beings have frequently tried to avoid knowingly leaving behind serious problems that will make the lives of future generations more difficult. In this case, technologically advanced societies are piling up an ever-growing stockpile of hazardous radioactive wastes that are likely to burden our descendants for *thousands of generations*.

Nuclear Accidents

Atmospheric bomb tests carried out from the 1940s through the 1970s diluted their considerable radioactive fallout by spreading it over a wide area. In contrast, the two most serious civilian nuclear power plant accidents to date released significant doses of harmful radiation over much smaller areas. The first such incident was on April 26, 1986 at Chernobyl in Soviet Ukraine, where a poorly designed, poorly operated, nuclear reactor overheated and blew up. In the aftermath, no general warning was issued to the public for two days. In that period, radioactive debris heavily blanketed the nearby town of Pripyat, where most of the plant's workers lived, as well as the surrounding countryside, before slowly spreading out over several European countries. Chaos and widespread population evacuations eventually followed, and as desperate efforts got underway to get the reactor under control, only limited observations could be made on the exact extent of the radioactive fallout.

The best estimates are that several thousand emergency workers received fatal doses of radiation—although most did not die immediately—and that roughly 10,000 additional people are likely to die of cancer as a result of the radioactive iodine-131, cesium-137 and other isotopes released. By 2005, an estimated 6,000 children and teenagers who were exposed to Chernobyl's fallout had gone on to develop thyroid cancer, and 15 children were known to have died of it within that same period.[23] At the time, authorities gave little thought—perhaps understandably—to the environmental damage underway around Chernobyl, and it was never adequately documented. But it is known that all the trees in a pine forest close to the reactor died, no doubt accompanied by widespread deaths in local animal populations. A bit farther away, the pine trees managed to survive, but went on to absorb large amounts of radiation in the years since the disaster. Should lightning or arson ever kindle any serious forest fires there, it seems likely that clouds of radioactive smoke would be released and spread serious contamination far outside the present exclusion zone.

The short, eight-day half-life of iodine-131 meant that most of what was released at Chernobyl decayed to stable xenon-131 within three months following the disaster. But the explosions at the reactor also spewed out significant amounts of radioactive cesium-137 and strontium-90, whose half-lives of 30.17 and 28.8 years meant that they could continue to cause harm for up to three centuries. The escape of these isotopes meant that dangerous, if slowly declining, contamination would remain in Ukraine, Belarus, Russia, and several other European countries into the twenty-third century. Even early in the twenty-first century, a fraction of animals that graze and browse outdoors in the region still contain levels of radiation that make them unfit for human consumption.[24]

The second greatest civilian nuclear power accident so far in terms of the radiation released occurred on March 11, 2011. On that date, a large undersea earthquake off the northeastern coast of Japan created an enormous tsunami that together with seismic tremors quickly killed an estimated 20,000 people. In addition to devastating a number of towns and villages, the tsunami flooded a large, coastal power plant at Fukushima housing six nuclear reactors, three of which were operating at the time. The gigantic wave of ocean water cut the power plant's supply of electricity from the regional grid, while also disabling the plant's emergency back-up generators. As a result, all three operating reactors overheated, releasing large amounts of radioactive steam into the air along with significant amounts of radioactive iodine and cesium. In the aftermath, large volumes of water carrying radioactive debris were also discharged both into the sea and onto the land.[25] As at Chernobyl, the most immediate hazard at Fukushima was posed by the radioiodine. On this occasion, however, better prepared authorities managed to successfully limit contamination from that dangerous, short-lived isotope. Significant amounts of radioactive cesium-134 (half-life of two years) and cesium-137 (half-life of 30 years) were also released, and these still account for the greatest contamination in the areas surrounding the reactor. Their long-term effects on people and on land and ocean ecosystems is not yet known. The undersea earthquake and the resulting tsunami could not have been avoided, but in July 2012 a committee appointed by the Japanese parliament to investigate the incident came to the conclusion that the meltdown and explosions had been entirely preventable. The reactors would not have overheated if all of the safety measures that were supposed to be in place at the plant had been working at the time of the disaster. They concluded that both the owners of the plant and the government agency responsible for regulating it had failed in their responsibilities to ensure safe operation.[26]

Beyond their main applications in nuclear weapons and power generation, the third notable use of radioactive isotopes has been in medicine, where very small quantities of several isotopes have proven useful in medical diagnosis and in the treatment of certain kinds of cancer. This is probably the area of nuclear technology that has provided the most favorable balance between benefits and costs, although it does produce very small quantities of radioactive waste which requires very careful handling and disposal. And even in this tightly controlled and well-administered field, a number of major and minor mishaps have still occurred over the years, especially in the management of retired equipment containing radioactive materials. By far the most serious of these occurred in Goiania in Brazil in 1987 when radioactive cesium-137 from abandoned radiotherapy equipment was released into the environment and then intentionally dispersed by people not aware of its true nature. In the incident's aftermath, 249 people were seriously contaminated, with twenty people developing full-blown radiation sickness, and four dying. In addition to the deaths and injuries, the incident in Goiania also required very costly and extensive de-contamination and clean-up.[27]

Over the relatively short history of nuclear technology, serious nuclear accidents have thankfully been uncommon, although minor mishaps and small radiation releases have occurred with some frequency.[28] Given the complexity of the technologies involved, the planned growth in civilian nuclear power plants, the unpredictability of natural disasters, and a certain unavoidable level of human error, the record suggests that additional nuclear mishaps are almost certain to occur in the future.

Radiation and Life

By the 1950s, atomic energy had been embraced by a number of industrialized countries, and in those nations, the creation and occasional release of potentially harmful radioisotopes had become regular occurrences. As a result, during the second half of the twentieth century, people and all other organisms experienced somewhat higher levels of background radiation. Radiation on earth has three sources: some streams in from space where it originates in stars and other celestial objects; some is produced by the breakdown of naturally occurring radioactive elements like uranium and thorium which make up a small fraction of earth's original rocks; the rest comes from human nuclear technology. In the past, the radiation arriving from space has varied in response to distant events, but it is not known to have reached high levels during recent evolutionary time. The natural radiation that is emitted by elements in the earth's crust has very gradually diminished over time as more and more of the originally radioactive atoms of elements like uranium have decayed into non-radioactive end products like lead. The radiation released by human technology has varied since the 1940s, peaking early on with atmospheric bomb explosions, and then continuing at lower levels ever since.[29]

Radiation brings with it the potential to harm living things because the energetic particles or waves that make it up are able to penetrate living cells. Sometimes, when this occurs, the radiation passes right through without causing any damage. At other times, however, the energy carried by the wave or particle interacts with sub-atomic particles inside the cell to cause harm. It may, for example, break chemical bonds and so deactivate or destroy key molecules. If the bonds disrupted happen to be those belonging to the cell's hereditary DNA, this can interfere with a cell's ability to reproduce, to repair itself, or to make essential cellular substances. Once it reaches a cell, radiation can also turn previously stable chemical structures into highly reactive chemical *ions* which can then interact destructively with other vital cellular molecules including DNA. For this reason dangerous radioactivity is sometimes referred to as *ionizing* radiation.[30]

Since background radiation and errors in copying DNA have been present ever since the beginning of life, natural selection long ago produced a complex set of enzymes and mechanisms designed to allow cells to repair different kinds of biochemical disruption, particularly damage to DNA. Without these internal safeguards, the frequency of cellular disease and cancer would be far higher than it is. Vertebrates add to this the ability of certain immune cells to identify and destroy some mutated cells before they can multiply to become full-blown cancers.

Earlier we saw that exposure to a number of chemicals, including benzene, can sometimes cause cancer (see Chapter 5). Radiation can also have this effect; the major difference in its action compared to carcinogenic chemicals is that radiation is more likely to cause random damage across a wide range of molecules. Chemical pollutants, in contrast, more commonly attack specific structures, and are also able to insert themselves physically into biological molecules where they can permanently interfere with their functioning. In either case, however, *it's the dose that makes the poison*, and once exposure to either carcinogenic chemicals or radiation reaches a certain level, natural mechanisms can no longer repair cellular molecules as fast as they are being damaged. When this happens, cells may die or enter a dormant state, *or* they may turn into cancer cells. Since cellular damage accumulates over time and has at least two different causes (for a few types of cancer, viruses may constitute a third cause), some cases of cancer seem to result from a combination of exposure to toxic chemicals *and* to ionizing radiation (an evolutionary perspective on the nature of cancer is offered in Chapter 5).

★　★　★

Nuclear technologies were one of the most remarkable inventions of the twentieth century, and in a few medical applications, they offered unique advantages at minimal cost and risk. The creation of nuclear weapons, on the other hand, must be viewed as a grave setback to the continuation of life on earth. Even if never again exploded, existing weapons stockpiles, and the wastes generated in their production, will be sources of dangerous and even deadly pollution for hundreds of thousands of years into the future. The human generations that created those materials will not be held in high regard by posterity.

As technologies go, the use of atomic energy to generate electricity seems to fall somewhere in between the clear negatives of nuclear weapons and the relative positives of nuclear medicine. Electric power itself seems, on balance, to have been a force for good in society. Despite some significant environmental costs, it has vastly improved people's standard of living, their comfort and safety, their access to information, their health, and the quality of their consumer products. Electricity has also been essential for the advancement of science and engineering, which together have created the basis for a new level of human existence. Yet current nuclear power plants, all in all, have too many problems to qualify as a basically sound way for humanity to meet a significant portion of its power needs. When viewed over their entire life cycle—from planning and construction right up to the time hundreds of thousands of years from now when the wastes they have generated will no longer be hazardous to life—they are far and away the most expensive of all possible sources of energy. Not only do they present significant short-term hazards—as the Fukushima disaster made clear—but the intractable problems of high level waste disposal involve risks and costs not associated with any other industry or power source. (Private investment has never been willing to assume all of these risks without government underwriting, a fact suggesting that nuclear power plants' dangers and potential liabilities are far greater than many governments are willing to publicly admit.)

In the future, advances in science and engineering may lead to at least partial solutions to the economic and safety challenges presented by the continuing flows of dangerous, radioactive wastes from nuclear electricity plants. But there is certainly no guarantee of that. And although such breakthroughs have been anticipated for many years, so far they have failed to materialize. In contrast, recent decades have produced striking progress in the development of far more life-friendly sources of electricity ranging from solar, wind and tidal power, to hybrid cars (which generate some free electricity from their brakes), as well as dramatic opportunities for energy conservation in the form of more efficient vehicles, better building insulation and more efficient lighting. In the near term, at least, it seems clear that it would be far better to scale back or even gradually phase out existing nuclear power plants, while aggressively replacing them with renewable energy supplies and greater energy efficiency. Life has many millions of years yet to run on planet earth. Human energy desires should not be allowed to curtail its future.

★　★　★

Renewable Energy

In the second half of the twentieth century, with population and energy demand both skyrocketing and early impacts of global warming becoming apparent, many people began to think about *renewable energy*. It was a simple idea. On earth, vast amounts of power are at work at any given time completely apart from human civilization. (Large, individual hurricanes, for example, are estimated to contain forces that are many times greater than all human electrical generating capacity.[31]) Most energy on earth comes from the sun which sends us large amounts of light and heat every day. This is the power that allows plants to grow and creates temperatures suitable for the delicate chemical reactions necessary for life. Only a tiny percentage of solar energy, however, is actually captured by green plants; most of it instead ends up as heat which warms the world and drives planet-wide circulations of air and water. Wind is the most noticeable component of this circulation, although huge amounts of solar power also drive earth's water cycle which includes evaporation, precipitation and the vast movements of seawater we know as ocean currents. Gravitational attraction of seawater by the moon adds additional rhythmic motion to the water's movement. All of these involve very large amounts of energy. In addition, a smaller flow of energy to the earth's surface comes from the intense volcanic heat continually present in earth's core which results from a combination of gravity and nuclear reactions. In a few places, like Iceland, it has proven possible to tap such *geothermal* energy to provide not only hot water and space heating but also power generation.

Hydro-power

As we saw at the beginning of Chapter 6, the power of water flowing downhill and of wind were the two earliest sources of energy available to people other than muscle power and the heat from fire. Of the two, rivers offered a more concentrated source of energy than wind, and by the third century BC, water-powered grain mills had begun to be constructed in Greece. Between then and the nineteenth century, tens of thousands of them were built around the world, becoming regular features of the rural countryside. In addition, some were adapted to other repetitive tasks like sawing wood or hammering metal.[32] In the mid-nineteenth century, as rotating electrical generators began to be built, it did not take long for people to wonder whether the power of flowing water could also be used to run them. By 1880, hydroelectric plants had begun to be built in a number of countries, first to produce DC power, and then AC, and by the end of the decade, hundreds were in operation around the world.[33]

Central to the eventual importance of hydro-power were important advances in water-wheel design made in 1848 by James Francis and Uriah Boyden, two engineers working in the United States. Together they developed turbines that were able to extract nearly 90 percent of the energy of the water flowing through them, a design so efficient that it is still in use today.[34] Francis, in particular, was a pioneering nineteenth-century hydraulic engineer whose "mathematical and graphical calculation methods improved turbine design and engineering. His analytical methods allowed confident design of high efficiency turbines to exactly match a site's water flow and water pressure (*water head*) conditions."[35] By employing the same quantitative, scientific approach used by other great nineteenth-century engineers like James Watt, Charles Parsons and Rudolph Diesel, Francis and Boyden managed to make important advances which helped launch the modern field of hydro-power engineering.

The combination of the Francis turbine with alternating current electrical systems (see Chapter 5) opened up an era of rapid growth for hydroelectricity. In some cases, its development went hand-in-hand with the creation of reservoirs intended to control flooding downstream, or in some places to provide irrigation water. Many hundreds of dams were built around the world, and in areas where large flows were available, hydroelectricity soon became an important component of many countries' electric supply, providing 40 percent of all US power needs, for example, by 1920.[36]

Hydro-power installations continued to be built throughout the twentieth century and into the twenty-first. By the 2010s, roughly 2,000 large power dams were in operation worldwide along with an even larger number of smaller ones.[37] In 2012 China opened the largest hydro plant ever built, the massive Three Gorges Dam on the Yangtze River. From the middle of the twentieth century on, however, growth in demand for electricity around the world became so great that hydro development could not keep up, and its share of supply in many countries fell throughout the late twentieth century. By the early twenty-first century, with many potential sites for hydro-power already built upon, hydro's percentage of the United States' electric supply, for example, had fallen to just 10 percent. Worldwide, it currently produces about 16 percent of all the electricity that is consumed.[38]

When widespread dam construction began in the early decades of the twentieth century, hydro-power seemed like a nearly perfect source of energy. It generated no significant air or water pollution while operating, and the supply of falling water to run it was provided free by earth's natural cycles. This made its power quite inexpensive once the costs of dam construction have been paid off. In subsequent decades, however, it became clear that hydro-power did have a number of significant impacts, the most important being damage to the natural river ecosystems whose waters were being tapped for energy.

Understanding the impact of dams requires some sense of what undammed river environments are like. All rivers are powered by the slope of land from higher to lower elevations and most end up at sea level where they meet the ocean in a specialized, brackish water environment known as an estuary. The flow of water in rivers often varies seasonally—increasing in the temperate zone spring following the melting of winter snow and in the tropics following rainy seasons—and decreasing in warmer, drier seasons. These cycles also lead to changes in water temperature, suspended materials and dissolved nutrients and gases. As rivers move along downslope, the tremendous weight and energy of the moving water carries along both mineral matter and living and dead biological material. Where streams flow rapidly, as in mountainous regions, this may result in nutrient-poor but oxygen-rich conditions, often with visible rocks and white water. Where the slope of the land is less, streams slow down but often carry more nutrients and silt and less dissolved oxygen. Over time many aquatic species have evolved a life pattern which matches the seasonal and local conditions of their home river system.

For many aquatic creatures, natural rivers are continuous ecosystems; it is easy to move downstream and not hard to move upstream as long as they can swim against the current or hitch a ride on a bird or migrating fish. The possibilities of such movements greatly increase the resilience of river ecosystems in several ways. They allow desirable genetic exchange among populations of creatures living in different locations along the river's course, and this, in turn, maintains variation and allows improved genetic combinations to spread. They permit replacement of local populations that may have been wiped out by locally severe conditions. And they allow some fish species including eels and salmon to successfully maintain life patterns that involve living alternately in the river and in the ocean at different stages of their life cycle, with a long migration joining the two.

So what are the environmental impacts when a dam is built? One of the very first things is that the enormous amounts of concrete that are generally used to build modern dams require a lot of cement, a material whose manufacture produces about one pound of carbon dioxide for every pound of cement, along with other air pollutants.[39] After construction, the normal rate of flow of the river is altered, usually with a raising of water levels behind the dam and a lowering immediately downstream. This leads to many changes in the river including its water temperature, silt load and dissolved gases, and this may, in turn, lead to the elimination of some species and an increase in the populations of others. Creatures all the way up the food chain are impacted by these changes. Once the dam is operating, control of river flows will mostly follow economics rather than natural cycles, and this is likely to cause on-going ecosystem disruption.

Dams make it hard for creatures to move downstream and nearly impossible for many to move upstream. Those moving down often have to go through the power turbines, which often spells death for larger species. Those moving up are confronted with the wall of the dam blocking their way. In some places, fish ladders have been built to allow upstream passage, but even the best of them excludes much of the pre-dam upstream traffic. The interruption of the natural movements of river creatures interferes with the genetics of the organisms inhabiting the river and also with their ability to recolonize upstream areas following declines. A series of dams, in effect, turns a single continuous body of water into a set of interconnected ponds with very limited movement in the upstream direction. For certain populations of long distance migrants like salmon and eels, dams may make it simply impossible for them to complete their natural life cycle.[40]

Besides their direct effects on living things, dams also have other impacts. Over time, much of the silt which the river used to transport downstream becomes trapped behind the dam where it tends to build up, instead of spreading out over many miles of flood plain or of deltas where the river enters the sea. Eventually, the reservoir may become shallower, reducing its ability to generate power. In some cases, farmland fertility downstream suffers from the loss of silt, and river deltas may begin to erode away because their soil is no longer being replenished. There may also be direct human costs. The construction of the massive Three Gorges Dam and the reservoir behind it displaced more than one and a quarter million people from their homes as numerous towns and villages were flooded. Its huge reservoir also drowned many sites of significant archaeological and cultural importance.[41] Overall, the generation of hydroelectricity from rivers seems to have provided a net benefit. But it has also come at significant cost to the natural world, especially to the rich river ecosystems that existed before the dams were built.

The damming of streams and rivers has long provided a relatively straight-forward way of obtaining useful energy, although that approach is not the only way that the power in moving water can be tapped for human purposes. A second approach is based on recovering energy from tidal flows. Even hundreds of years ago, in a number of locations in Europe and North America where large variations in water level occurred between low and high tides, dams and grist mills were built on rocky coves, creating salt ponds from which falling seawater could provide energy to grind grain for a number of hours each day. The principle was simple: high tide filled a pond with incoming salt water over a spillway, and then once the tide was out, a gate would gradually let the water out in such a way as to spin a wheel that was mechanically linked to grindstones. For a number of reasons, including the constantly changing hour of the tides, there were serious limitations to mills like this, yet quite a few were nonetheless built and operated successfully.[42]

In the twentieth century, the idea of tidal power was revived not with the notion of turning a water wheel for mechanical power, but rather of driving a turbine to generate electricity. The first such plant began operation in France in 1966 near the mouth of the Rance River, and even today it continues to generate modest amounts of power.[43] Only a handful of similar plants have been built around the world, but newer technologies are currently under development which may allow electricity to be produced from dispersed turbines or other devices that are tethered in coastal waters and do not require the construction of tidal ponds or dams.[44]

Wind Energy

Useful harnessing of wind energy dates back thousands of years in sailboats, and over a millennium in mills dedicated to grinding grain or pumping water. In nineteenth- and early-twentieth-century America, for example, the use of rotating wind machines to mechanically pump water up from wells for both livestock and human use was extremely common. Millions of such units are believed to have been in operation by the 1930s.[45] Wind power for the generation of electricity, however, mostly came later, with a few experimental wind turbines being set up to drive generators from the 1880s on. The wind, however, does not blow all the time, and it is not always available during periods of high electrical demand. Right from the start this led to the use of batteries to store some of the wind-generated power. In the nineteenth and twentieth centuries however, the kinds of batteries available proved to be a weak link in the system due to their expense and inefficiency. Only in Denmark was a serious attempt made to connect electricity-generating wind machines directly to the main power grid.

From the 1980s on, however, people once again became interested in wind-generated electricity as a promising source of renewable energy, and this time they managed to develop it into a successful technology. The key changes were that now much larger wind machines could be built, and their power could be fed directly into the electrical grid for immediate use wherever it was needed. In addition, very advanced engineering borrowed from the aviation industry, coupled with computer aided design and computerized controls, was employed to create sophisticated, efficient, durable turbines. In recent decades, these have rapidly increased in size to provide ever-increasing amounts of power, with the largest unit's blades towering higher than a 70-story building and capable of generating eight million watts.[46]

As with other natural resources, usable winds are not distributed uniformly around the earth. The strongest air currents are miles up in the atmosphere (e.g., the jet streams) or in remote locations in Antarctica or on high mountains, and no good way has yet been found to extract useful energy from them. Despite these limitations, wind power grew substantially in recent decades and now generates about 4 percent of the world's electricity, a share that is expected to increase in coming years. Its use, however, is spread very unevenly around the world, with some countries generating none, and Denmark, a wind power leader, meeting about forty percent of its power needs from wind with plans to significantly expand its use in coming decades.[47] Depending on the local availability of wind versus fossil fuels or photo-voltaic, wind-generated electricity can be more or less expensive than other forms of power generation. Wind turbines produce essentially no pollution once they are installed. If located on land, their main negative environmental impacts are their appearance, the sounds they produce, and the relatively modest numbers of birds and bats killed in collisions with their turbine blades. Turbines located in coastal waters have the same potential problems of appearance and bird mortality; their foundations may interfere with swimming organisms; and the process of building them can sometimes disturb marine ecosystems.[48] Despite these limited drawbacks, their development represents a very important step forward in developing completely renewable sources of electricity that are highly compatible with the long-term survival of the rest of the living world.

Solar Energy

Hydroelectric power and wind energy are the two leading sources of renewable electricity today. But breakthroughs in physics, materials science, and manufacturing during the nineteenth and twentieth centuries have also created a third important source of renewable energy. In 1887 Heinrich Hertz, a German physicist, showed that light shining on certain materials could produce a small electric current, an observation that was repeated by other experimenters. In 1905, Albert Einstein published a theoretical analysis of this phenomenon which by then had come to be called the *photoelectric effect*. Einstein's work proved to be of great theoretical importance, although in the early twentieth century it did not immediately produce any important practical applications.[49]

After World War II, however, tremendous advances in materials science produced a significant breakthrough. In 1954, researchers at Bell Labs in the United States demonstrated silicon crystals that were capable of transforming 4.5 percent of the solar energy falling on them into electricity. By 1957, samples capable of 8 percent efficiency were developed, and by 1960 14 percent.[50] These levels were initially achieved at great cost, using unusual materials in demonstration projects, and by the second decade of the twenty-first century, the rarified descendants of those early research cells were able to transform fully 44 percent of sunlight into electricity.[51]

For practical, cost-effective energy production, however, the exotic materials and complex production techniques needed to produce such high levels of energy capture were simply not workable. Instead, industry settled on a material called *polycrystalline silicon* which still dominates photovoltaic technology today. Manufactured at reasonable cost, panels made from it typically reach efficiencies of around 15 percent.[52] Photovoltaic units have no moving parts and require little if any maintenance—mostly keeping the dust and snow off—and normally last twenty to thirty years outdoors. The manufacture of solar-electric panels does require significant quantities of water and electricity along with some toxic chemicals, and it takes about two years for a given panel to generate as much electricity as it took to make it. Solar arrays also require land if they are mounted on the ground and like wind turbines have a visual impact. But other than that, they are marvelous, renewable power-generating devices which deserve to be much more widely installed.

During the twentieth century, relatively high cost and low production volume of panels prevented photovoltaics from becoming an important source of energy in national power grids. Instead they were used primarily for specialized applications, most notably as sources of power in remote locations. In the twenty-first century, however, this slowly began to change, and by 2010 photovoltaic installations had begun to grow faster than other sources of renewable power; by 2013, the installed worldwide capacity of more than 100 gigawatts was roughly 20 times as great as it had been just eight years earlier.[53] (A gigawatt equals one billion watts, or roughly enough electricity to run about 660,000 1500 watt microwave ovens at the same time. Installed photovoltaic capacity of 100 gigawatts in 2013 was one hundred times greater than this, or enough to run 66 million microwave ovens.[54])

Direct production of electric power from the sun using flat solar panels is a truly remarkable technology even though more than three quarters of the solar energy falling on photovoltaic panels is still either reflected or thrown off as waste heat. There are, however, two other solar technologies that are able to capture somewhat higher percentages of incident solar energy. The first of these uses the sun's heat to boil water which is then used to drive either a steam turbine or a Stirling engine connected to a generator. The idea is simple, although the implementation is somewhat complex. Curved mirrors are set up on a sunny site and arranged so as to concentrate solar energy using one of two different basic designs. The first uses parabola shaped troughs which focus the sun's heat on tubes filled with molten salts. These get very hot, and are then pumped to a boiler to make steam. The other incorporates large arrays of mirrors which focus all their heat directly on a central tower containing a boiler. So far, the parabolic reflector design has been more widely implemented, most notably in Spain and in the American southwest. In regions where the sun shines most of the time such *concentrated solar power* represents a workable technology which is likely to be more widely used in the future. Concentrated solar is capable of reaching significantly higher efficiencies than current mainstream photovoltaics (greater than 30%), but the power plants utilizing it are much more complicated than solar panels, and it is not really suited for small or dispersed installations. Each mirror in a concentrating plant requires two motors to track the sun during the day, hot circulating salts in the case of parabolic trough designs, and all the complexities of a steam generating plant to actually produce the electricity.[55]

In contrast to the engineering complexities of concentrated solar-electric, the third effective solar technology—direct solar heating of water or air—is relatively simple and efficient, although it can provide only heat and not electricity. That the sun provides warmth is an everyday observation, and just as many animals like to bask in the sun, for thousands of years people in cool climates have oriented their houses to maximize the solar heat they receive. In the nineteenth century, as window glass became more and more affordable, it also became feasible to use larger and larger sun-facing windows to bring solar warmth into buildings. By the mid-twentieth century, the addition of effective insulation made it possible to build houses which could get a lot of their heating directly from the sun. Large expanses of windows are not appropriate for every building, however, and so devices designed to generate solar-heated air were also developed for roof and wall mounting. Modern versions of such *solar hot air* collectors feature dark surfaces that can capture more than 90 percent of the sun's energy, and then transfer most of it inside to warm a building.

It is also possible to heat water in this way. From the nineteenth century on, people realized that an outdoor water tank painted black and exposed to the sun could produce warm water. Later, it was found that enclosing the water tank in a glassed-in box facing the sun could create hot water. Over the course of the twentieth century, such devices were repeatedly improved to the point where they became a practical, economical technology, especially in warm and sunny climates. Several different effective designs were developed, and over the last forty years tens of millions of small and medium-sized units have been employed around the world, with the largest number being installed in China, and the highest percentage of use occurring in Israel. Many use a highly efficient technology based on heat collecting pipes surrounded by evacuated glass tubes—like a transparent insulating flask.[56]

As is frequently the case with solar hot air and photovoltaic, solar hot water heaters can only provide part of a house's energy needs; and they are not very effective in climates that are both cold and cloudy. But anything that displaces a significant fraction of nuclear or fossil-fuel derived energy represents an enormously positive step forward towards a sustainable world. And in the sunnier parts of the world, solar energy can often meet essentially all of a home's needs for both heat and electricity. (With additional equipment, solar energy can also be used to drive an air conditioning unit to provide cooling.)

Energy Efficiency and Conservation

So far, we have dwelled on renewable energy technologies that can provide electricity or heat without continual consumption of fossil fuels or nuclear energy. In broader perspective, however, it will also be important to continue finding ways to carry out important activities using *less energy* than in the past while also shifting lifestyles away from the most energy-intensive pastimes.

The kind of changes needed are not new. Increasing energy efficiency has been typical of advancing human know-how—think of the dramatic reductions in fuel consumption as James Watt's steam engine replaced the original Thomas Newcomen design, or when the Parsons steam turbine replaced reciprocating steam engines for generating electricity. Many other areas of technology have seen parallel efficiency gains. Houses and offices are now much better insulated and also have improved lighting provided by fluorescent or light-emitting diode (LED) lights in place of the older inefficient incandescents. Early mass-produced, low-powered automobiles like Ford's Motel T went about 17 miles on a gallon of gasoline; roughly a century later, the significantly more powerful third generation Toyota Prius Hybrid, introduced in 2010, went three times as far on the same amount of fuel, and at higher speeds.[57]

One area in which important improvements in efficiency from combustion-based power generation have not yet fully been realized is through the *cogeneration* of electricity and heat. Power plants intended to make only electricity can rarely utilize more than 40 percent of the total heat energy produced by their fuel. If, however, they are also designed to pipe the waste steam and hot water they produce to nearby buildings or factories for heating and hot water use, their thermal efficiency can be much higher. By the late twentieth century, the pairing of such cogeneration plants with specialized absorption chillers even made it possible for a single power plant to provide electricity, steam heat, and cold water for air conditioning. And this meant that as much as 80 percent of all the energy consumed by the plant could be captured for useful purposes.[58] In New York City, for example, more than three million customers currently receive both electricity and heat produced by a series of large cogeneration plants, while similar services are in place on a smaller scale in many parts of Europe. Initial investment costs for such installations are relatively high due to the necessary network of pipes under streets. But with very high efficiencies, long-term costs can be competitive, while environmental impacts are dramatically reduced.[59]

Conclusion

Recent decades have seen significant progress in energy efficiency and alternative energy sources. If current best practices and technologies were followed everywhere, the sum of energy-related problems—including pollution from excess carbon dioxide—would be much reduced. But there are many barriers to implementing these improvements. First there must be a strong social desire to reduce environmental impacts, and this in turn requires both widespread public understanding *and* endorsement of these goals by government policy-makers. In countries like the United States, however, where important groups within society explicitly reject scientific understanding of environmental problems like global warming, it is hard for government to effectively regulate or encourage change. And in developing countries, where poverty and related ills are either still present or have only just been left behind, it is difficult to rally support for significant investments whose pay-offs will partly benefit other nations or generations yet unborn.

Over the past several decades, even as technology has steadily advanced, people have failed to come into balance with the rest of the living world because our numbers and our standards of living have grown even faster than our progress in alternative energy and energy efficiency. New power-hungry technologies have continually appeared—from motor vehicles to jet aircraft, from central heating to air conditioning, from outdoor lighting to internet server farms—and ever more people consuming more and more power has easily outstripped the otherwise impressive gains made in efficiency.[60] For this reason, total pollution has continued to increase even though we now create less pollution for each unit of energy produced. Going forward, we will need to have not only more sources of alternative energy (along with more green chemistry and green manufacturing), but also more awareness on the part of both government and the public of what kinds of actions damage the environment. We will also need a cultural reorientation away from power-hungry and wasteful activities towards satisfying, lower-impact pastimes.

The advance of civilization has been closely tied to the harnessing of different kinds of energy, starting with human and animal muscle power. Human labor, and even that of *beasts of burden*, however, had very real limits, and most agricultural societies limited to those two sources of energy eventually tended to experience land shortages, poverty or recurring episodes of famine. Capturing the energy of flowing water for mills and of wind for sailboats certainly led to improvements in human life, although it was not until the first taming of steam in 1712 that the real potential of artificial sources of power was glimpsed. In the three centuries since then, the human energy economy has grown millions of times over, supporting unprecedented population expansion and dramatic improvements in daily life. For humanity as a whole, this may have been good. Yet during most of those 300 years, the true cost of these same energy technologies was never reckoned. And when, in the late twentieth century, this was done for the first time, vast and damaging impacts were discovered.

The continuation of human civilization is now tied tightly to the use of high levels of energy. If we are to ensure a safe, healthy, future for our descendants and for the rest of the living world, we must start by recognizing the serious damage that has been caused by many existing power technologies. And once we have done that, we must commit ourselves to replacing destructive energy supplies with sources of power that truly are sustainable.

Notes

1   IAEA, Pioneering nuclear science: the discovery of nuclear fission, www.iaea.org/newscenter/news/2013/nuclfission.html; American Institute of Physics, The discovery of fission, www.aip.org/history/exhibits/mod/fission/fission1/01.html.
2   Nobelprize.org, The development and proliferation of nuclear weapons, www.nobelprize.org/educational/peace/nuclear_weapons/readmore.html.
3   World Nuclear Association, Hiroshima, Nagasaki, and subsequent weapons testing, http://world-nuclear.org/info/inf52.html.
4   World Nuclear Association, Plutonium, www.world-nuclear.org/information-library/nuclear-fuel-cycle/fuel-recycling/plutonium.aspx.
5   Wikipedia, Nuclear weapons testing, http://en.wikipedia.org/wiki/Nuclear_weapons_testing.
6   World Nuclear Association, Nuclear-powered ships, www.world-nuclear.org/information-library/non-power-nuclear-applications/transport/nuclear-powered-ships.aspx.
7   US Department of Energy, The history of nuclear energy, http://energy.gov/sites/prod/files/The%20History%20of%20Nuclear%20Energy_0.pdf.
8   Popular Mechanics, 1954, Britain's pioneer atomic-power plant, *Popular Mechanics* (June): 74, https://books.google.com/books?id=1t4DAAAAMBAJ&pg=PA74&dq=1954+Popular+Mechanics+January&hl=en&sa=X&ei=VTQqT6rRIMbo2AX_39SHDw#v=onepage&q&f=true.
9   IAEA, The database on nuclear power reactors, www.iaea.org/pris.
10 Purdue University Bodner Research Web, Nuclear fission and nuclear fusion, http://chemed.chem.purdue.edu/genchem/topicreview/bp/ch23/fission.php; Wikipedia, Nuclear fission product, https://en.wikipedia.org/wiki/Nuclear_fission_product.
11 L. Sherwood et al., 2005, *Animal Physiology*, Brooks/Cole; Radioactivity.eu.com, Iodine-131, www.laradioactivite.com/en/site/pages/Iodine_131.htm.
12 S. L. Simon et al., 2006, Fallout from nuclear weapons tests and cancer risk, *American Scientist* 94(1): 48–57.
13 L. Sherwood et al., 2005, *Animal Physiology*, Brooks/Cole.
14 USEPA, Radionuclide basics: cesium-137, www.epa.gov/radiation/radionuclide-basics-cesium-137; C. Wessells, 2012, Cesium-137: a deadly hazard, http://large.stanford.edu/courses/2012/ph241/wessells1; US Centers for Disease Control and Prevention Radioisotope Brief, Cesium-137 (Cs-1), http://emergency.cdc.gov/radiation/isotopes/cesium.asp.
15 US Department of Energy, Final Bomb Design, www.osti.gov/opennet/manhattan-project-history/Events/1942-1945/final_design.htm.
16 World Nuclear Association, Plutonium, www.world-nuclear.org/information-library/nuclear-fuel-cycle/fuel-recycling/plutonium.aspx; International Panel on Fissile Materials, Fissile material stocks, www.fissilematerials.org.
17 World Nuclear Association, Plutonium, www.world-nuclear.org/info/Nuclear-Fuel-Cycle/Fuel-Recycling/Plutonium.
18 US Agency for Toxic Substances and Disease Registry, Toxicological profile for plutonium, www.atsdr.cdc.gov/toxprofiles/tp143.pdf; G. L. Voelz, 2000, Plutonium and

health, *Los Alamos Science* 26: 74–89, www.fas.org/sgp/othergov/doe/lanl/pubs/00818013.pdf.

19 Wikipedia, Nuclear power, https://en.wikipedia.org/wiki/Nuclear_power.

20 World Nuclear Association, Plans for new reactors worldwide, http://world-nuclear.org/information-library/current-and-future-generation/plans-for-new-reactors-worldwide.aspx.

21 IAEA, 2007, Estimation of global inventories of radioactive waste and other radioactive materials, www-pub.iaea.org/MTCD/publications/PDF/te_1591_web.pdf.

22 R. L. Brathwaite and T. Christie, 2014, Mineral commodity report 14: Gold, www.researchgate.net/publication/253564570_Mineral_Commodity_Report_14_-_Gold.

23 US Nuclear Regulatory Commission, Backgrounder on Chernobyl nuclear power plant accident, www.nrc.gov/reading-rm/doc-collections/fact-sheets/chernobyl-bg.html.

24 Wikipedia, Chernobyl disaster, https://en.wikipedia.org/wiki/Chernobyl_disaster.

25 F. N. von Hippel, 2011, The radiological and psychological consequences of the Fukushima Daiichi accident, *Bulletin of the Atomic Scientists* 67(5): 27–36.

26 K. Mahr, 2012, Report: Fukushima nuclear disaster was man-made, *Time* (July 5), http://world.time.com/2012/07/05/report-fukushima-nuclear-disaster-was-manmade.

27 IAEA, 1988, The radiological accident in Goiania, www-pub.iaea.org/mtcd/publications/pdf/pub815_web.pdf.

28 Wikipedia, Nuclear and radiation accidents and incidents, https://en.wikipedia.org/wiki/Nuclear_and_radiation_accidents_and_incidents.

29 P. A. Karam and S. A. Leslie, 2005, Changes in terrestrial natural radiation sources over the history of life, *Radioactivity in the Environment* 7: 107–117.

30 Purdue University-Bodner Research Web, Ionizing Radiation, http://chemed.chem.purdue.edu/genchem/topicreview/bp/ch23/radiation.php.

31 NOAA, Hurricane Research Division, How much energy does a hurricane release?, www.aoml.noaa.gov/hrd/tcfaq/D7.html.

32 R. D. Hansen, Water wheels, http://waterhistory.org/histories/waterwheels; Wikipedia, Water wheel, https://en.wikipedia.org/wiki/Water_wheel.

33 Wikipedia, Hydroelectricity, https://en.wikipedia.org/wiki/Hydroelectricity.

34 US National Park Service, James B. Francis, www.nps.gov/lowe/learn/historyculture/upload/JB%20Francis_%20Lowell%20Notes.pdf.

35 Wikipedia, Francis turbine, https://en.wikipedia.org/wiki/Francis_turbine.

36 Wikipedia, Hydroelectricity, https://en.wikipedia.org/wiki/Hydroelectricity.

37 International Hydropower Association, Maps, www.hydropower.org/maps.

38 International Energy Agency, Hydropower Essentials, www.iea.org/publications/freepublications/publication/Hydropower_Essentials.pdf.

39 M. Rubenstein, 2012, Emissions from the Cement Industry Columbia University-Earth Institute, http://blogs.ei.columbia.edu/2012/05/09/emissions-from-the-cement-industry.

40 Foundation for Water and Energy Education, How a hydroelectric project can affect a river, http://fwee.org/environment/how-a-hydroelectric-project-can-affect-a-river.

41 Wikipedia, Three Gorges Dam, https://en.wikipedia.org/wiki/Three_Gorges_Dam.

42 W. E. Minchinton, 1979, Early tide mills: some problems, *Technology and Culture* 20(4): 777–786.

43 British-Hydro.org, La Rance tidal power plant, www.british-hydro.org/downloads/La%20Rance-BHA-Oct%202009.pdf.

44 Alternative Energy News, Tidal power, www.alternative-energy-news.info/technology/hydro/tidal-power.

45 US Department of Energy, History of wind energy, www.energy.gov/eere/wind/history-wind-energy.

46 Wikipedia, History of wind power, https://en.wikipedia.org/wiki/History_of_wind_power.

47 Denmark.dk, A world-leader in wind energy, http://denmark.dk/en/green-living/wind-energy.

48 Wikipedia, Environmental impact of wind power, https://en.wikipedia.org/wiki/Environmental_impact_of_wind_power.

49 G. Elert, Photoelectric effect, http://physics.info/photoelectric.

50 M. Bellis, 2014, History: Photovoltaics Timeline, http://inventors.about.com/od/timelines/a/Photovoltaics.htm.

51 Phys.org, World record solar cell with 44.7% efficiency, http://phys.org/news/2013-09-world-solar-cell-efficiency.html.

52 Energy Informative, The homeowner's guide to solar panels, http://energyinformative.org/best-solar-panel-monocrystalline-polycrystalline-thin-film/#polycrystalline.

53 Union of Concerned Scientists, 2014, Solar power on the rise, www.ucsusa.org/clean_energy/our-energy-choices/renewable-energy/solar-power-technologies-and-policies.html.

54 US Department of Energy, Estimating appliance and home electronic energy use, http://energy.gov/energysaver/articles/estimating-appliance-and-home-electronic-energy-use.

55 National Renewable Energy Laboratory, Concentrating solar power research, www.nrel.gov/csp.

56 Energy Star, How it works—solar water heaters, www.energystar.gov/products/water_heaters/water_heater_solar/how_it_works; Wikipedia, Solar water heating, https://en.wikipedia.org/wiki/Solar_water_heating.

57 Ford Motor Company, Model T facts, https://media.ford.com/content/fordmedia/fna/us/en/news/2013/08/05/model-t-facts.html; US Department of Energy, Compare side-by-side, www.fueleconomy.gov/feg/Find.do?action=sbs&id= 31767.

58 US Energy Information Administration, What is the efficiency of different types of power plants?, www.eia.gov/tools/faqs/faq.cfm?id=107&t=3; GHD, Cogeneration plants increase energy efficiency, www.ghd.com/global/projects/cogeneration-plants; GE Power Generation, Combined heat and power, https://powergen.gepower.com/applications/chp.html.

59 European Association for the Promotion of Cogeneration, What is co-generation?, www.cogeneurope.eu/what-is-cogeneration_19.html.

60 G. Glanz, 2012, Power, pollution and the internet, *New York Times* (September 22), www.nytimes.com/2012/09/23/technology/data-centers-waste-vast-amounts-of-energy-belying-industry-image.html?_r=0.

INDEX

Printed and bound by CPI Group (UK) Ltd, Croydon, CR0 4YY

22/10/2024

01777606-0012